Lecture Notes in Computational Science and Engineering

17

Editors
M. Griebel, Bonn
D. E. Keyes, Norfolk
R. M. Nieminen, Espoo
D. Roose, Leuven
T. Schlick, New York

T0223610

Springer
Berlin
Heidelberg
New York
Barcelona
Hong Kong
London
Milan
Paris
Singapore
Tokyo

Barbara I. Wohlmuth

Discretization Methods and Iterative Solvers Based on Domain Decomposition

With 82 Figures and 25 Tables

 Springer

Barbara I. Wohlmuth
Institut für Mathematik
Universität Augsburg
Universitätsstraße 14
86159 Augsburg, Germany
e-mail: Barbara.Wohlmuth@math.uni-augsburg.de

Cataloging-in-Publication Data applied for

Die Deutsche Bibliothek - CIP-Einheitsaufnahme

Wohlmuth, Barbara:
Discretization methods and iterative solvers based on domain
decomposition / Barbara I. Wohlmuth. - Berlin ; Heidelberg ; New York;
Barcelona ; Hong Kong ; London ; Milan ; Paris ; Singapore ; Tokyo :
Springer, 2001
 (Lecture notes in computational science and engineering ; 17)
 ISBN 3-540-41083-X

Mathematics Subject Classification (1991): 65N30, 65N15, 65F10, 65N55, 65N50

ISSN 1439-7358
ISBN 3-540-41083-X Springer-Verlag Berlin Heidelberg New York

Springer-Verlag Berlin Heidelberg New York
a member of BertelsmannSpringer Science+Business Media GmbH

© Springer-Verlag Berlin Heidelberg 2001
Printed in Germany

Cover Design: Friedhelm Steinen-Broo, Estudio Calamar, Spain
Cover production: *design & production* GmbH, Heidelberg
Typeset by the author using a Springer TeX macro package

Printed on acid-free paper   SPIN 10725050   46/3142/LK – 5 4 3 2 1 0

Preface

DOMAINE: [dɔmɛn]
Ce domaine est encore fermé aux savants

DÉCOMPOSER: [dekɔ̃poze]
Décomposer un problème pour mieux le résoudre

*Micro Robert: Dictionnaire du français primordial

The numerical approximation of partial differential equations, very often, is a challenging task. Many such problems of practical interest can only be solved by means of modern supercomputers. However, the efficiency of the simulation depends strongly on the use of special numerical algorithms. Domain decomposition methods provide powerful tools for the numerical approximation of partial differential equations arising in the modeling of many interesting applications in science and engineering. Although the first domain decomposition techniques were used successfully more than hundred years ago, these methods are relatively new for the numerical approximation of partial differential equations. The possibilities of high performance computations and the interest in large-scale problems have led to an increased research activity in the field of domain decomposition.

However, the meaning of the term "domain decomposition" depends strongly on the context. It can refer to optimal discretization techniques for the underlying problems, or to efficient iterative solvers for the arising large systems of equations, or to parallelization techniques. In many modern simulation codes, different aspects of domain decomposition techniques come into play, and the overall efficiency depends on a smooth interaction between these different components. The coupling of different discretization schemes, the coupling of different physical models, and many efficient preconditioners for the algebraic systems can be analyzed within an abstract framework. At first glance these aspects seem to be rather independent. However, all have one central idea in common: The decomposition of the underlying global problem into suitable subproblems of smaller complexity. In general, a complete decoupling of the global problem into many independent subproblems, which are easy to solve, is not possible. Since, the subproblems are very often coupled, there has to be communication between the different subproblems. Although the term optimal depends on the context, the proper handling of the information transfer across the interfaces between the subproblems is of major importance for the design of optimal methods. In the case of discretization techniques, a priori estimates for the discretization errors have to be considered. They very much depend on the appropriate couplings across the interfaces which are often realized by matching conditions. The jump across the interfaces which measures the nonconformity of the method has to be bounded in a suitable way. In the case of iterative solvers, the convergence rate and the computational effort for one iteration step measure the quality of a method. To obtain scalable iteration schemes, very often, one has to include a suitable global problem of small complexity.

In this work, both discretization techniques and iterative solvers are addressed. A brief overview of different approaches is given and new techniques and ideas are proposed. An abstract framework for domain decomposition methods is presented and an analysis is carried out for new techniques of special interest. Optimal estimates for the methods considered are established and numerical results confirm the theoretical predictions.

Chapter 1 concerns special discretization methods based on domain decomposition techniques. In particular, the decomposition of geometrical complex structures into subdomains of simple shape is of special interest. Another example is the decomposition into substructures on which different physical models are relevant. Then, for each of these subproblems, an optimal approximation scheme involving the choice of the triangulation as well as the discretization can be chosen. However to obtain optimal discretizations for the global problem, the discrete subproblems have been glued together appropriately. Here, we focus on mortar finite element methods.

To start, we review the standard mortar setting for the coupling of Lagrangian conforming finite elements in Sect. 1.1. Both standard mortar formulations – the nonconforming positive definite problem and the saddle point problem based on the unconstrained product space – are given.

In Sect. 1.2, we introduce and analyze alternative Lagrange multiplier spaces. We derive abstract conditions on the Lagrange multiplier spaces such that the nonconforming discretization schemes obtained yield optimal a priori results. Lagrange multiplier spaces based on a dual basis are of special interest. In such a case, a biorthogonality relation between the nodal basis functions of these spaces and the finite element trace spaces holds. A main advantage of these new Lagrange multiplier spaces is that the locality of the support of the nodal basis functions of the constrained space can be preserved.

With this observation in mind, we introduce a new equivalent mortar formulation defined on the unconstrained product space in Sect. 1.3. We show that the non-symmetric formulation can be analyzed as a Dirichlet–Neumann coupling. Based on the elimination of the Lagrange multiplier, we derive a symmetric positive definite formulation on the unconstrained product space, and the equivalence to the positive definite problem on the constrained space is shown. Two formulations, a variational as well as an algebraic one, are presented and discussed. A standard nodal basis for the unconstrained product space can be used in the implementation. The stiffness matrix associated with our new variational form can be obtained from the standard one on the unconstrained space by local operations.

Section 1.4 concerns two examples of non-standard mortar situations. Each of them reflects an interesting feature of the abstract general framework, and illustrates the flexibility of the method. We start with the coupling of two different discretization schemes. The matching at the interface is based on the dual role of Dirichlet and Neumann boundary conditions. Two different equivalent formulations are given for the coupling of mixed and standard conforming finite elements. In our second example, we rewrite the nonconforming Crouzeix–Raviart finite elements as mortar finite elements. We consider the extreme case that the decomposition of the domain is given by the fine triangulation and that therefore the number of subdomains tends to infinity as the discretization parameter of the triangulation tends to zero.

Finally in Sect. 1.5, we present several series of numerical results. In particular, we study the influence of the choice of the Lagrange multiplier space on the discretization errors. Examples with several crosspoints, a corner singularity, discontinuous coefficients, a rotating geometry, and a linear elasticity problem are considered. A second test series concerns the influence of the choice of the non-mortar side. Adaptive and uniform refinement techniques are applied. In our last test series, we consider the influence of jumps in the coefficient on an adaptive refinement process at the interface.

Chapter 2 concerns iterative solution techniques based on domain decomposition. A brief overview of general Schwarz methods, including multigrid techniques, is given in Sect. 2.1. Examples for the standard H^1-case illustrate overlapping, non-overlapping, and hierarchical decomposition techniques. The following sections contain new results on non-standard situations; we discuss vector field discretizations as well as mortar methods.

Section 2.2 focuses on an iterative substructuring and a hierarchical basis method for Raviart–Thomas finite elements in 3D. We start with the definition of the local spaces and the relevant bilinear forms and subspaces. The central result of this section is established in Subsect. 2.2.2; it is a polylogarithmical bound independent of the jumps of the coefficients across the subdomain boundaries of our iterative substructuring method. The technical tools are discussed in detail with particular emphasis on the role of trace theorems, harmonic extensions, and dual norms applied to finite element spaces. As in the 2D case for standard Lagrangian finite elements, we introduce three different types of subspaces called V_H, V_F, and V_T. We cannot avoid the use of a global space to obtain quasi-optimal bounds. But in contrast to the standard Lagrangian finite elements in 3D, the low dimensional Raviart–Thomas space associated with the macro-triangulation formed by the subregions can be used to obtain quasi-optimal results where the constant does not depend on the jumps of the coefficients across the subdomain boundaries.

Sections 2.3–2.5 concern different iterative solvers for mortar finite element formulations. In Sect. 2.3, we combine the idea of dual basis functions for the Lagrange multiplier space with standard multigrid techniques for symmetric positive definite systems. The new mortar formulation, analyzed in Sect. 1.3, is the point of departure for the introduction of our iterative solver. We define and analyze our multigrid method in terms of level dependent bilinear forms, modified transfer operators, and a special class of smoothers which includes a standard Gauß–Seidel smoother. Convergence rates independent of the number of refinement steps are established for the W-cycle provided that the number of smoothing steps is large enough. The numerical results confirm the theory. Moreover asymptotically constant convergence rates are obtained for the V-cycle with one pre- and one postsmoothing step.

Section 2.4 concerns a Dirichlet–Neumann type algorithm for the mortar method. It turns out to be a block Gauß–Seidel solver for the unsymmetric mortar formulation on the product space. Numerical results illustrate the influence of the choice of the damping parameter. The transfer of the boundary values at the interface is realized in terms of a scaled mass matrix. This matrix is sparse if and only if dual Lagrange multiplier spaces are used.

In Sect. 2.5, we study a multigrid method for the saddle point formulation. Two different types of smoothers are discussed; a block diagonal and one reflecting the saddle point structure. In the second case, the exact solution of the modified Schur complement system is replaced by an iteration, resulting in an inner and an outer iteration. This multigrid method is given for the standard mortar formulation as presented in Sect. 1.1. In contrast to the two previous sections, the use of dual Lagrange multiplier spaces does, in general, not reduce the computational costs for one iteration step.

Acknowledgments: It is a great pleasure for me to thank my colleagues, friends, and parents for their support. In particular, I wish to thank Prof. Ronald H.W. Hoppe, University of Augsburg, for his support throughout this work, Prof. Yuri A. Kuznetsov, University of Houston, who introduced me to work in this area, and Prof. Dietrich Braess, Ruhr–University Bochum, for his interest in my work since we first met at the annual GAMM meeting in Braunschweig, 1994. This work was supported in part by a Habilitandenstipendium of the Deutsche Forschungsgemeinschaft and in part by the National Science Foundation under Grant NSF-CCR-9732208. In particular, I would like to thank my coworkers, Rolf Krause, PhD Andrea Toselli and Priv.-Doz. Christian Wieners for their support and fruitful discussions. Finally, it is a great pleasure for me to thank Prof. Olof B. Widlund of the Courant Institute, New York University. His criticism, his encouragement, and his preference for simplicity influenced my work.

Augsburg, October 2000   *Barbara Irmgard Wohlmuth*

Contents

1. Discretization Techniques Based on Domain Decomposition

This chapter concerns domain decomposition methods as discretization techniques for partial differential equations. We present different approaches within the framework of mortar methods [BMP93, BMP94]. Originally introduced as a domain decomposition method for the coupling of spectral elements, these techniques are used in a large class of nonconforming situations. Thus, the coupling of different physical models, discretization schemes, or non-matching triangulations along interior interfaces of the domain can be analyzed by mortar methods. These domain decomposition techniques provide a more flexible approach than standard conforming formulations. They are of special interest for time dependent problems, rotating geometries, diffusion coefficients with jumps, problems with local anisotropies, corner singularities, and when different terms dominate in different regions of the simulation domain. Very often heterogeneous problems can be decomposed into homogeneous subproblems for which efficient discretization techniques are available. To obtain a stable and optimal discretization scheme for the global problem, the information transfer and the communication between the subdomains is of crucial importance; see Fig. 1.1.

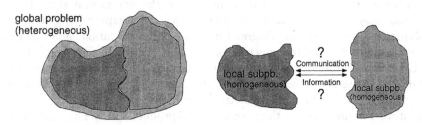

Fig. 1.1. Decomposition of a global problem into homogeneous subproblems

One major requirement is that the interface between the different regions is handled appropriately. Very often suitable matching conditions at the interfaces are formulated as weak continuity conditions. One possibility is to use a dual set of boundary conditions. Then, the coupling can be realized in terms of a Lagrange multiplier. The analysis of the resulting jump terms across the interfaces plays an essential role for the a priori estimates of the discretization schemes. In particular, optimal methods can only be obtained if the consistency error is small enough compared with the best approximation error on the different subdomains. The consistency error measures the nonconformity of the method and controls the information transfer at the interface.

In Sect. 1.1, we review the definition of a standard mortar finite element method. For an overview of mortar techniques including spectral elements, we refer to [BD98, Ben99, BM97, BMP93, BMP94, LSV94]. Both standard mortar formulations – the nonconforming positive definite problem and the saddle point problem based on the unconstrained product space – are given. Basic ideas and techniques are explained in a standard context. In the following sections, new ideas resulting in alternative spaces are studied, and different modifications and aspects of the methods are considered. We also introduce a new positive definite mortar formulation based on the unconstrained product space.

In Sect. 1.2, a general framework for the construction of new Lagrange multiplier spaces is given. Sufficient conditions on the Lagrange multiplier space are introduced to obtain an optimal order global method. So far the Lagrange multiplier spaces have been defined as finite element trace spaces. We focus on a special type of Lagrange multiplier spaces based on a dual basis. This idea is new and simplifies the implementation as well as the iterative solution of the arising problem. The nodal basis functions of the Lagrange multiplier and the trace spaces satisfy a biorthogonality relation. As a consequence the constraints at the interfaces can be eliminated locally, and the locality of the supports of the basis functions of the constrained space can be guaranteed.

We focus on the introduction of a new equivalent mortar formulation based on the unconstrained product space in Sect. 1.3. The biorthogonality relation yields a diagonal mass matrix on the non-mortar side. Then, the Lagrange multiplier can be obtained in a local postprocessing step, involving the inverse of that diagonal matrix. This observation serves as our starting point for the introduction of our new mortar formulation on the unconstrained product space. In a first step, we show that the resulting unsymmetric form is nothing else than a Dirichlet–Neumann coupling between the different subdomains. A symmetrized form gives rise to a positive definite variational formulation on the unconstrained product space. This third mortar formulation can be implemented much easier and solved more efficiently than the standard nonconforming and saddle point formulations.

Section 1.4 concerns several nonconforming situations which are analyzed within the framework of mortar finite element methods. The examples illustrate the large variety and flexibility of mortar techniques. A first example is given in Subsect. 1.4.1. The coupling of primal and dual methods is analyzed, using the duality of essential and natural boundary conditions as our point of departure. Two different formulations are given for this coupling, including a saddle point problem where the Lagrange multiplier space is defined by piecewise constants. The second example is based on Crouzeix–Raviart elements. In contrast to the previous example, the number of subdomains tends to infinity when the meshsize tends to zero. Each element is regarded as a subdomain and the Lagrange multiplier space is the product space of one dimensional spaces associated with the edges of the triangulation.

Finally in Sect. 1.5, numerical results are given which illustrate the performance of mortar discretization techniques. We start with a comparison of the discretization errors for the different choices of Lagrange multiplier spaces given in Sect. 1.2. In a second part, we focus on the choice of the mortar side in case of discontinuous coefficients and highly nonconforming triangulations. Finally, the use of a posteriori error estimators shows that two completely different situations arise at the interfaces depending on the choice of the mortar side.

Throughout this chapter, we emphasize the role of stable projections for the best approximation property of the constrained space, the approximation property of the Lagrange multiplier space for the consistency error, discrete inf-sup conditions for the a priori estimates for the Lagrange multiplier, mesh dependent norms for measuring the nonconformity, and dual basis functions as Lagrange multipliers for the locality of the support of the nodal basis functions of the constrained space.

The following elliptic second order boundary value problem

$$-\operatorname{div}(a\nabla u) + bu = f \quad \text{in } \Omega \ , \tag{1.1}$$
$$u = 0 \quad \text{on } \partial\Omega \ ,$$

will serve as our model problem. Here, a is a uniformly positive definite matrix, $a_{ij} \in L^{\infty}(\Omega)$, $1 \le i,j \le d$, $f \in L^{2}(\Omega)$, $0 \le b \in L^{\infty}(\Omega)$, and $\Omega \subset \mathbb{R}^{d}$, $d = 2,3$, is a bounded polygonal domain. For simplicity, we assume that the coefficients a_{ij} and b are constant on each element of the triangulations. Although the detailed analysis is given in 2D, most of our results also hold in 3D. In the following, we only point out differences between the 2D and 3D case if different techniques for the proofs are required or if different qualitative results are obtained. In those cases, the analysis of the 3D case is carried out separately. All constants $0 < c \le C < \infty$ throughout this work are generic and might depend on the coefficients a and b, the aspect ratio of the elements and subdomains, and the order of the discretization method but they do not depend on the meshsize.

1.1 Introduction to Mortar Finite Element Methods

In this section, we briefly review the standard mortar method for the coupling of Lagrangian finite elements. We recall the nonconforming positive definite formulation as well as the saddle point problem and the a priori estimates.

An examination of the mortar projection shows that the support of a basis function on the non-mortar side is, in general, non-local.

Let Ω be decomposed into K non-overlapping polyhedral subdomains Ω_k such that

$$\overline{\Omega} = \bigcup_{k=1}^{K} \overline{\Omega}_k \ .$$

We restrict ourselves to the geometrical conforming situation where the intersection between the boundaries of any two different subdomains $\partial\Omega_l \cap \partial\Omega_k$, $k \neq l$, is either empty, a vertex, a common edge or face in 3D; see Fig. 1.2. We call it an interface only in the latter case.

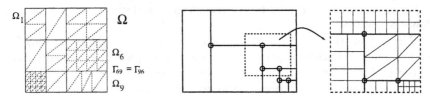

Fig. 1.2. Geometrically conforming (left) and nonconforming (right) situation

Geometrically nonconforming situations are technically more difficult to handle. A possibility to reduce these complications is to require that each vertex of the decomposition is also a vertex of each adjacent triangulation; see the right part of Fig. 1.2.

We define for each subdomain a simplicial triangulation $\mathcal{T}_{k;h_k}$, the meshsize of which is bounded by h_k. The finite element space of conforming P_{n_k}-elements on Ω_k associated with $\mathcal{T}_{k;h_k}$, $n_k \geq 1$, which satisfy homogeneous Dirichlet boundary conditions on $\partial\Omega \cap \partial\Omega_k$, is denoted by $X_{h_k;n_k}$. No boundary conditions are imposed on $X_{h_k;n_k}$ in the case that $\partial\Omega \cap \partial\Omega_k = \emptyset$. The results can be easily generalized to other types of triangulations; see Fig. 1.3.

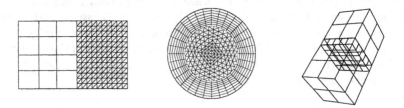

Fig. 1.3. Decomposition and non-matching triangulations in 2D and 3D

Each interface $\partial\Omega_l \cap \partial\Omega_k$ is associated with a $(d-1)$-dimensional triangulation, inherited from either $\mathcal{T}_{k;h_k}$ or $\mathcal{T}_{l;h_l}$. In general, these triangulations do not coincide. The interfaces are denoted by γ_m, $1 \leq m \leq M$. For each interface, there exists a couple $1 \leq l < k \leq K$ such that $\overline{\gamma}_m = \partial\Omega_l \cap \partial\Omega_k$. Since we are working with finite dimensional Lagrange multiplier spaces on the interfaces, we have to define appropriate triangulations on γ_m. The triangulation on γ_m is called $\mathcal{S}_{m;h_m}$, and its elements are boundary edges in 2D and boundary faces in 3D of either $\mathcal{T}_{l;h_l}$ or $\mathcal{T}_{k;h_k}$. The choice is arbitrary but should be fixed. Then, by definition, the Lagrange multiplier space inherits its triangulation from the non-mortar side. The adjacent side is called the mortar side. We denote the subdomain associated with the non-mortar side by $\Omega_{n(m)}$ and the one associated with the mortar side by $\Omega_{\bar{n}(m)}$.

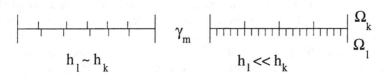

Fig. 1.4. Different non-matching triangulations at an interface in 2D

In general $\mathcal{T}_{k;h_k}$ and $\mathcal{T}_{l;h_l}$ do not match across the common interface. We remark that no conditions on the triangulations are imposed at the common interfaces. Figure 1.4 shows two characteristic situations for non-matching triangulations at the interface. The situation on the left typically arises for time dependent problems, e.g., in the case of sliding meshes. In the case of highly discontinuous coefficients a, global triangulations having a jump in the meshsizes on the different subdomains might yield better results; see the right of Fig. 1.4. For both situations, the mortar finite element method yields optimal a priori bounds for the discretization error in the energy norm [BMP93, BMP94]. The constant in the a priori estimate in 2D depends neither on the ratio of the meshsize of two adjacent subdomains nor on the distortion. In contrast to the 2D case, the ratio $h_{\mathrm{mor}}/h_{\mathrm{non}}$ enters, in general, into the a priori estimates in 3D; see, e.g., [BM97, BD98]. Here, h_{mor} denotes the meshsize on the mortar side and h_{non} the one on the non-mortar side. Under some additional assumptions on the triangulation, the factor $h_{\mathrm{mor}}/h_{\mathrm{non}}$ can be avoided. We refer to [KLPV00, Woh00a] for a more precise analysis of the constants in the a priori estimates and to Subsect. 1.5.3 for some numerical results illustrating the influence of the choice of the non-mortar side. In the rest of this section, we restrict ourselves to the 2D case. The discussion of the 3D case is included in Sect. 1.2.

To obtain the mortar approximation u_h, as a solution of a discrete variational problem, there are so far two main approaches. The first one has been introduced in [BMP93, BMP94] and gives rise to a positive definite nonconforming variational problem. It is defined on a subspace V_h of the product space, the elements of which satisfy weak continuity conditions across the interfaces. The constrained finite element space V_h is given by

$$V_h := \{ v \in L^2(\Omega) \mid v_{|\Omega_k} \in X_{h_k;n_k}, \, 1 \le k \le K \, ,$$
$$\int_{\gamma_m} [v] \mu \, d\sigma = 0, \, \mu \in M_{h_m}(\gamma_m), \, 1 \le m \le M \} \, ,$$

where the test space $M_{h_m}(\gamma_m)$ is given by

$$M_{h_m}(\gamma_m) := \{ \mu \in L^2(\gamma_m) \mid \mu = w_{|\gamma_m}, \, w \in X_{h_{n(m)};n_{n(m)}} \, ,$$
$$\mu_{|e} \in P_{n_m - 1}(e), \text{ if } e \in \mathcal{S}_{m;h_m} \text{ contains an endpoint of } \gamma_m \} \, ,$$
(1.2)

where $n_m := n_{n(m)}$. The indices l and k are reserved for the subdomains whereas the index m is used for an interface, and we have to understand the definition of n_m in this sense. In 3D, the definition of the Lagrange multiplier space has to be modified in the neighborhood of $\partial \gamma_m$. We remark that $M_{h_m}(\gamma_m)$ is a modified trace space of codimension two, associated with the 1D triangulation on the non-mortar side which is inherited from $\mathcal{T}_{n(m);h_{n(m)}}$. Thus, the space $M_{h_m}(\gamma_m)$ depends on the choice of the non-mortar side. The global product $\prod_{m=1}^{M} M_{h_m}(\gamma_m)$ is denoted by M_h, and is a subspace of $L^2(\mathcal{S})$, where $\overline{\mathcal{S}} := \cup_{m=1}^{M} \overline{\gamma}_m$.

In principle, it is also possible to introduce a new independent 1D triangulation on each interface γ_m which is inherited neither from $\mathcal{T}_{n(m);h_{n(m)}}$ nor from $\mathcal{T}_{\bar{n}(m);h_{\bar{n}(m)}}$. Then, we have to impose suitable conditions on the mesh-size or add adequate bubble functions to the finite element space to obtain a discrete inf-sup condition. Such stabilization techniques are discussed in [BM00, BFMR98] for three-field approaches.

The nonconforming formulation of the mortar method can be given in terms of the constrained space V_h: Find $u_h \in V_h$ such that

$$a(u_h, v_h) = (f, v_h)_0, \quad v_h \in V_h \, ; \tag{1.3}$$

see [BMP93, BMP94]. Here, the bilinear form $a(\cdot, \cdot)$ is defined as

$$a(v, w) := \sum_{k=1}^{K} \int_{\Omega_k} a \nabla v \cdot \nabla w + bv \, w \, dx, \quad v, w \in \prod_{k=1}^{K} H^1(\Omega_k) \, .$$

We remark that continuity was imposed at the vertices of the decomposition in the first papers about mortar methods. However, this condition can be removed without loss of stability. Both these settings guarantee uniform ellipticity of the bilinear form $a(\cdot, \cdot)$ on $V_h \times V_h$, as well as a best approximation error and a consistency error of optimal order; see [BMP93, BMP94]. Combining Lax–Milgram's and Strang's Lemmas, it can be shown that a unique solution of (1.3) exists. If the weak solution u of (1.1) is smooth enough and H^2-regularity holds, then we have the following a priori estimates for the discretization error in the broken H^1-norm and in the L^2-norm

$$\|u - u_h\|_1^2 \le C \sum_{k=1}^{K} h_k^{2n_k} |u|_{n_k+1;\Omega_k}^2 \ ,$$

$$\|u - u_h\|_0^2 \le C \, h^2 \sum_{k=1}^{K} h_k^{2n_k} |u|_{n_k+1;\Omega_k}^2 \ ;$$

(1.4)

see [BMP93, BMP94, BDW99]. Here, we use a standard Sobolev notation for norms which are not explicitly defined and set $\|v\|_1^2 := \sum_{k=1}^{K} \|v\|_{1;\Omega_k}^2$, $v \in \prod_{k=1}^{K} H^1(\Omega_k)$. $\|\cdot\|_{s;D}$ stands for the H^s-norm on the open set $D \subset \Omega$, and $|\cdot|_{s;D}$ is the corresponding semi norm. In the case that $D = \Omega$, the index Ω is suppressed.

Furthermore, we have ellipticity of the bilinear form $a(\cdot,\cdot)$ on $Y \times Y$

$$a(v,v) \ge c \, \|v\|_1^2, \quad v \in Y \ ,$$

(1.5)

where Y is defined as

$$Y := \left\{ v \in \prod_{k=1}^{K} H^1(\Omega_k) \mid \ v_{|\partial\Omega} = 0, \int_{\gamma_m} [v] \, d\sigma = 0, 1 \le m \le M \right\} \ ;$$

(1.6)

see [BM95]. We note that $V_h \subset Y$. Then, the energy norm $\|v\|^2 := a(v,v)$, $v \in Y$ is equivalent to the broken H^1-norm. In [Gop99, Theorem IV.1], it has been established that the ellipticity constant is independent of the number of subdomains. A similar estimate is given for the three field approach in [BM00]. The proof of the a priori bounds for the discretization error (1.4) is based on a best approximation error result and an a priori estimate for the consistency error; see [BMP93, BMP94]. We note that very often the a priori estimate in the broken H^1-norm is given in the weaker form

$$\|u - u_h\|_1 \le C \sum_{k=1}^{K} h_k^{n_k} |u|_{n_k+1;\Omega_k} \ .$$

However, in [Woh00a] it has been shown that (1.4) also holds with a constant independent of the number of subdomains.

The constraints at the interfaces guarantee that the consistency error is at least as good as the sum of the best approximations errors on the different subdomains. Replacing V_h in (1.3) by the unconstrained product space

$$X_h := \{ v \in L^2(\Omega) \mid \ v_{|\Omega_k} \in X_{h_k;n_k}, 1 \le k \le K \}$$

(1.7)

yields a consistency error that is not bounded in terms of the meshsize.

To prove the best approximation error of V_h, the mortar projection, $\Pi_m : C(\gamma_m) \longrightarrow W_{h_m}(\gamma_m)$, plays an important role. Here, the trace space $W_{h_m}(\gamma_m)$ is given by

$$W_{h_m}(\gamma_m) := \{ \mu \in C(\gamma_m) \mid \ \mu = w_{|\gamma_m}, w \in X_{h_{n(m)};n_{n(m)}} \} \ ,$$

and Π_m is defined in terms of the Lagrange multiplier space

$$\Pi_m v - v \in C_0(\gamma_m), \quad \int_{\gamma_m} (v - \Pi_m v)\mu \, d\sigma = 0, \quad \mu \in M_{h_m}(\gamma_m) \ . \qquad (1.8)$$

We have just recalled the definition of the mortar projection as in the original mortar papers. In Subsect. 1.2.1, we will consider a more natural way in defining it. It can be easily seen that the operator Π_m is well defined; see [BMP93, BMP94]. For the analysis of the approximation error, it is sufficient to show that the mortar projection is uniformly stable in suitable norms. The H_0^1-stability of the mortar projection is proved in [BMP93, BMP94]. This operator can be extended in a stable way to a linear and continuous operator from $H_{00}^{1/2}(\gamma_m)$ onto $W_{0;h_m}(\gamma_m) := H_0^1(\gamma_m) \cap W_{h_m}(\gamma_m)$; we will still denote the extended operator by Π_m.

We find the following inclusions $X_h \cap H_0^1(\Omega) \subset V_h \subset X_h$ and observe that replacing V_h in (1.3) by $X_h \cap H_0^1(\Omega)$ or X_h does not provide a good discretization scheme. It is obvious that the quality of the nonconforming approach (1.3) and the properties of V_h depend on the space M_h. Let us consider the structure of V_h in more detail. In general, V_h is not a subspace of $H_0^1(\Omega)$ and thus (1.3) is a nonconforming finite element method. Even for a nested sequence of global triangulations, the corresponding finite element spaces are non nested. Secondly in general, no basis of V_h with local support can be constructed. The constraints are given in terms of a L^2-orthogonality of the jumps, and an element $v_h \in X_h$ belongs to V_h if and only if

$$\int_{\gamma_m} \Pi_m([v_h])\mu_h \, d\sigma = 0, \quad \mu_h \in M_{h_m}(\gamma_m), \ 1 \leq m \leq M \ .$$

In particular, if $v_h \in V_h$ and $[v_h] \in C_0(\gamma_m)$, $1 \leq m \leq M$, then $\Pi_m([v_h]) = 0$, $1 \leq m \leq M$. A nodal basis function on the mortar side has to be extended to the non-mortar side such that the matching conditions are satisfied.

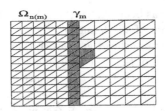

Fig. 1.5. Structure of the support of a nodal basis function in V_h, (standard)

Figure 1.5 shows the typical support of a basis function in V_h associated with an interface, where the non-mortar side is on the left. The support of such a nodal basis function on the non-mortar side is a strip of length $|\gamma_m|$ and width h_m, and the locality of the basis functions is lost.

Figure 1.6 illustrates the trace of a basis function of V_h on the two different sides of an interface for the two choices of the mortar side. In the left part of Fig. 1.6, the mortar side is associated with the finer triangulation whereas in the right part, it is associated with the coarser mesh. Although the basis functions on the non-mortar side have a global support, their values decrease exponentially.

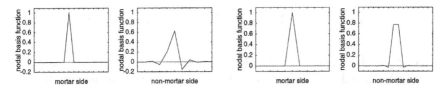

Fig. 1.6. Nodal basis function on a mortar and non-mortar side, (standard)

These observations motivate a second approach which was introduced in [Ben99] and further studied in [Woh99a]. The matching conditions on the interfaces are not imposed on the global space but realized by means of Lagrange multipliers. The starting point is a constrained minimization problem leading to the following saddle point formulation: Find $(u_h, \lambda_h) \in (X_h, M_h)$ such that

$$
\begin{aligned}
a(u_h, v) + b(v, \lambda_h) &= (f, v)_0, & v \in X_h \ , \\
b(u_h, \mu) &= 0, & \mu \in M_h \ ,
\end{aligned}
\tag{1.9}
$$

where the bilinear form $b(\cdot, \cdot)$ is given by the duality pairing on the interfaces

$$
b(v, \mu) := \sum_{m=1}^{M} \langle [v], \mu \rangle_{\gamma_m}, \quad v \in \prod_{k=1}^{K} H^1(\Omega_k), \ \mu \in \prod_{m=1}^{M} \left(H^{\frac{1}{2}}(\gamma_m) \right)'
$$

and $[v] := v|_{\Omega_{n(m)}} - v|_{\Omega_{\bar{n}(m)}}$. Here, $(H^{1/2}(\gamma_m))'$ denotes the dual space of $H^{1/2}(\gamma_m)$, and $\langle \cdot, \cdot \rangle_{\gamma_m}$ stands for the duality pairing.

Since the solution of the positive definite variational problem (1.3) and the first solution component of the saddle point problem (1.9) are equal, we use the same notation. The discrete Lagrange multiplier λ_h approximates the flux. A priori estimates for the error $\lambda - \lambda_h$ can be obtained, using the approximation property of M_h and a suitable inf-sup condition. Here, $\lambda|_{\gamma_m} := a \frac{\partial u}{\partial \mathbf{n}}$, where $\mathbf{n}$ is the outer normal on $\Omega_{n(m)}$. This issue was first addressed in [Ben99], where a priori estimates in the $H_{00}^{1/2}$-dual norm have been established. The $H_{00}^{1/2}$-dual norm is defined by

$$
\|\mu\|^2_{(H_{00}^{\frac{1}{2}}(S))'} := \sum_{m=1}^{M} \|\mu\|^2_{(H_{00}^{\frac{1}{2}}(\gamma_m))'} := \sum_{m=1}^{M} \sup_{v \in H_{00}^{\frac{1}{2}}(\gamma_m)} \frac{\langle v, \mu \rangle^2_{\gamma_m}}{\|v\|^2_{H_{00}^{\frac{1}{2}}(\gamma_m)}} \ ,
$$

where $\mu \in (H_{00}^{1/2}(S))'$ and $(H_{00}^{1/2}(S))' := \prod_{m=1}^{M}(H_{00}^{1/2}(\gamma_m))'$. The space $H_{00}^{1/2}(\gamma_m)$ can be interpreted as an interpolation space between $L^2(\gamma_m)$ and $H_0^1(\gamma_m)$.

Working with a posteriori estimates, it is often more convenient to deal with mesh dependent norms. Here, we consider a mesh dependent L^2-norm given by

$$\|\mu\|_{h-\frac{1}{2};S}^2 := \sum_{m=1}^{M} \sum_{e \in S_{m;h_m}} h_e \|\mu\|_{0;e}^2, \quad \mu \in L^2(S) \ ,$$

where h_e is the diameter of the element e; see [AT95]. A priori bounds for this mesh dependent L^2-norm are derived in [Woh99a]. The quality of the a priori estimates in the $H_{00}^{1/2}$-dual and this mesh dependent L^2-norm is the same.

We use $\| \cdot \|_{M'}$ to mean either the $H_{00}^{1/2}$-dual norm, $\| \cdot \|_{(H_{00}^{1/2}(S))'}$, or the mesh dependent L^2-norm, $\| \cdot \|_{h-1/2;S}$. As in the general saddle-point approach; see, e.g., [BF91], the essential point is to establish adequate inf-sup conditions; such bounds have been established with constants independent of the meshsize for both these norms; see [Ben99, Woh99a]. Thus, there exists a constant such that

$$\inf_{\substack{\mu_h \in M_h \\ \mu_h \neq 0}} \sup_{\substack{v_h \in X_h \\ v_h \neq 0}} \frac{b(v_h, \mu_h)}{\|\mu_h\|_{M'}\|v_h\|_1} \geq c \ . \tag{1.10}$$

We note that, so far, no inf-sup condition of the form (1.10) has been established for the $H^{1/2}$-dual norm, and that no a priori estimates are available in that norm.

If the solution u is regular enough, we find the following a priori estimate for the Lagrange multiplier by means of (1.10) and the approximation property of M_h

$$\|\lambda - \lambda_h\|_{M'}^2 \leq C \sum_{k=1}^{K} h_k^{2n_k} |u|_{1+n_k;\Omega_k}^2 \ . \tag{1.11}$$

The proof for the dual norm is given in [Ben99] and that for the mesh dependent L^2-norm in [Woh99a]. We remark that the bilinear form $b(\cdot, \cdot)$, defined on $X_h \times M_h$ is not uniformly continuous. To see this, we consider the following example. Let $v_h \in X_h$ be constant one on one subdomain Ω_{k_0} and zero elsewhere, and $\mu_h \in M_h$ be constant one on one interface $\gamma_{m_0} \subset \partial\Omega_{k_0}$ and zero elsewhere. Then, if the triangulation on the non-mortar side of γ_m is quasi-uniform with meshsize h_{m_0}, we find

$$| b(v_h, \mu_h) | \geq \frac{c}{\sqrt{h_{m_0}}} \|v_h\|_1 \|\mu_h\|_{h-\frac{1}{2};S} \ . \tag{1.12}$$

However, $b(\cdot, \cdot)$ is uniformly continuous for both $\|\cdot\|_{M'}$-norms if X_h is replaced by a suitable subspace. For the proof of (1.11), it is important that (1.10) also holds if the supremum is taken over this subspace. Details are worked out in Subsect. 1.2.3.

1.2 Mortar Methods with Alternative Lagrange Multiplier Spaces

In Sect. 1.1, the constraints at the interfaces γ_m are realized by means of a global L^2-projection, with test and trial spaces almost the same. In the nonconforming variational formulation (1.3), we have to face the problem that the basis functions of V_h cannot be given easily as linear combinations of those of X_h, since the construction of a basis of V_h involves the solution of mass matrix systems for each interface. Furthermore, the supports of the basis functions associated with the interfaces are non-local on the non-mortar sides. Following the second approach, (1.9), we can work with the unconstrained product space X_h. However, this formulation gives rise to an indefinite problem. We note that efficient iterative solvers for saddle point problems are often more complex than those for positive definite problems. In [SS98], it is shown that suitable lower dimensional Lagrange multiplier spaces also yield optimal discretization schemes and that, without loss of optimality, the order of the Lagrange multiplier space can be reduced by one compared with the standard approach given in Sect. 1.1. One characteristic of the Lagrange multiplier spaces introduced in [SS98] is that they are modified trace spaces of lower order conforming finite element discretizations.

Here, we propose different, more flexible, spaces for the Lagrange multiplier, in particular, spaces based on a dual basis. We recall that the Lagrange multiplier λ_h in the saddle point approach provides an approximation of the flux. A function $v \in H^1(\Omega_k)$ has its trace in $H^{1/2}(\partial \Omega_k)$ and the normal component of its flux is in the dual space $H^{-1/2}(\partial \Omega_k)$. This observation is the starting point for the construction of a new type of discrete Lagrange multiplier spaces. A special example of a dual basis in the first order case, $n_k = 1$, in 2D has already been studied in [Woh00a]. Here, we present dual basis spaces for the quadratic case in 2D and the first order case in 3D. Before we define our new dual bases, we develop a general framework for Lagrange multiplier spaces which yield optimal results.

We introduce a subspace of $L^2(\gamma_m)$ of dimension $N_m \leq \dim W_{0;h_m}(\gamma_m)$, and give appropriate assumptions under which this space can replace the standard Lagrange multiplier space defined in (1.2). The spaces $W_{h_m}(\gamma_m)$ and $W_{0;h_m}(\gamma_m) = W_{h_m}(\gamma_m) \cap H_0^1(\gamma_m)$ are introduced in Sect. 1.1 as finite element trace spaces on the non-mortar side. For convenience, we keep the same notations as before, and denote each element of the new abstract class of Lagrange multiplier spaces by $M_{h_m}(\gamma_m)$. We assume that there exists a basis $\{\psi_i \mid 1 \leq i \leq N_m\}$ of $M_{h_m}(\gamma_m)$ satisfying the following properties:

(Sa) Locality of the support: $\#(\text{supp}\psi_i) \leq C, \quad 1 \leq i \leq N_m$,

$$\#(p) \leq C, \quad p \in \gamma_m \ ,$$

where $\#(\text{supp}\psi_i)$ is the number of elements in $\mathcal{S}_{m;h_m}$ having a non-empty intersection with the simply connected support of ψ_i, and $\#(p)$ is the number of functions ψ_i such that the point p is contained in the support of ψ_i.

(Sb) Approximation property of $M_{h_m}(\gamma_m)$: For each $\mu \in H^{n_m - 1/2}(\gamma_m)$, there exists a $\mu_\psi \in M_{h_m}(\gamma_m)$ such that

$$\sum_{e \in \mathcal{S}_{m;h_m}} h_e \|\mu - \mu_\psi\|_{0;e}^2 \leq C h_m^{2n_m} |\mu|_{n_m - 1/2;\gamma_m}^2 \ .$$

In 2D, h_e is the length of the edge e and in 3D, h_e denotes the diameter of the face e. For a given basis, condition (Sa) is easy to verify and natural in the finite element context. Assumption (Sb) requires that the constants are contained in the Lagrange multiplier space. As in the standard mortar situation, a L^2-projection-like operator plays an essential role in establishing the approximation property for the constrained space. We base the more general mortar projection on a second set of linearly independent functions $\theta_i \in W_{0;h_m}(\gamma_m)$, $1 \leq i \leq N_m$, having simply connected local supports. The space

$$\widetilde{W}_{0;h_m}(\gamma_m) := \text{span} \{\theta_i \mid 1 \leq i \leq N_m\}$$

is a subspace of the trace space $W_{0;h_m}(\gamma_m)$. It is a proper subspace if $N_m < \dim W_{0;h_m}(\gamma_m)$. Obviously the space $\widetilde{W}_{0;h_m}(\gamma_m)$ cannot satisfy an approximation property for $H^1(\gamma_m)$. To establish an optimal order upper bound for the consistency error, we use the following modification

$$\widetilde{W}_{h_m}(\gamma_m) := \text{span} \{\widetilde{\theta}_i \mid 1 \leq i \leq N_m\}$$

where $\widetilde{\theta}_i \in W_{h_m}(\gamma_m)$ form a set of linear independent functions having simply connected local supports. By construction, the three spaces $M_{h_m}(\gamma_m)$, $\widetilde{W}_{0;h_m}(\gamma_m)$ and $\widetilde{W}_{h_m}(\gamma_m)$ have the same dimension.

The following two assumptions concern the discrete spaces $\widetilde{W}_{0;h_m}(\gamma_m)$ and $\widetilde{W}_{h_m}(\gamma_m)$ and their relation to $M_{h_m}(\gamma_m)$: We assume that for $M_{h_m}(\gamma_m)$, there exist two sets of basis functions defining $\widetilde{W}_{0;h_m}(\gamma_m)$ and $\widetilde{W}_{h_m}(\gamma_m)$ such that (Sc) and (Sd) holds:

(Sc) Approximation property of $\widetilde{W}_{0;h_m}(\gamma_m)$ and $\widetilde{W}_{h_m}(\gamma_m)$: For each $v \in H_0^s(\gamma_m)$ and $H^s(\gamma_m)$, $0 \leq s \leq 1$, there exists a $v_\theta \in \widetilde{W}_{0;h_m}(\gamma_m)$ and $\widetilde{W}_{h_m}(\gamma_m)$, respectively, such that

$$\sum_{e \in \mathcal{S}_{m;h_m}} \frac{1}{h_e^{2s}} \|v - v_\theta\|_{0;e}^2 \leq C |v|_{s;\gamma_m}^2 \ ,$$

$$|v_\theta|_{1;\gamma_m} \leq C |v|_{1;\gamma_m} \ , \quad v \in H^1(\gamma_m) \ .$$

(Sd) Spectral equivalence: $c\,D_\psi \le M_i D_i^{-1} M_i^T \le C\,D_\psi$,

$$c\,\tilde{D}_\psi \le M_i \tilde{D}_i^{-1} M_i^T \le C\,\tilde{D}_\psi \ , \quad i \in \{1,2\} \ ,$$

where the elements of the diagonal matrices D_1, D_ψ, $\tilde{D}_1$ and $\tilde{D}_\psi \in \mathbb{R}^{N_m \times N_m}$ are given by

$$(d_1)_{ij} := \delta_{ij} \|\theta_i\|_{0;\gamma_m}^2, \qquad (d_\psi)_{ij} := \delta_{ij} \|\psi_i\|_{0;\gamma_m}^2, \qquad 1 \le i,j \le N_m \ ,$$

$$(\tilde{d}_1)_{ij} := \delta_{ij} h_{\theta_i}^{-2} \|\theta_i\|_{0;\gamma_m}^2, \qquad (\tilde{d}_\psi)_{ij} := \delta_{ij} h_{\psi_i}^2 \|\psi_i\|_{0;\gamma_m}^2, \qquad 1 \le i,j \le N_m \ ,$$

where h_{θ_i} and h_{ψ_i} is the diameter of supp θ_i and supp ψ_i, respectively, and the mass matrix $M_1 \in \mathbb{R}^{N_m \times N_m}$ is defined by

$$(m_1)_{ij} := \int_{\gamma_m} \psi_i \theta_j \, d\sigma, \quad 1 \le i,j \le N_m \ .$$

To obtain D_2, $\tilde{D}_2$, and M_2, we replace the basis functions θ_i of $\widetilde{W}_{0;h_m}(\gamma_m)$, in the definitions of D_1, $\tilde{D}_1$, and M_1, by the basis functions $\tilde{\theta}_i$ of $\widetilde{W}_{h_m}(\gamma_m)$. The mass matrices M_i, $i \in \{1,2\}$, are sparse due to the locality of the supports. If $N_m = \dim W_{0;h_m}(\gamma_m)$, the approximation property (Sc) for $\widetilde{W}_{0;h_m}(\gamma_m)$ is automatically satisfied.

Remark 1.1. *The two approximation properties (Sb) and (Sc) are of different nature. (Sc) is a low order approximation property which does not depend on the order of the finite element approximation on the subspaces. In contrast, the order of the approximation property (Sb) depends on the order of the finite element approximation on the non-mortar side.*

We now use as our new Lagrange multiplier space $M_h = \prod_{m=1}^M M_{h_m}(\gamma_m)$, where $M_{h_m}(\gamma_m)$ is the abstract discrete space given by

$$M_{h_m}(\gamma_m) = \text{span}\{\psi_i \mid 1 \le i \le N_m\} \ .$$

The alternative nonconforming finite element space V_h is defined as before in terms of M_h

$$V_h := \big\{ v \in L^2(\Omega) \mid v_{|\Omega_k} \in X_{h_k;n_k}, \ 1 \le k \le K \ , \tag{1.13}$$
$$\textstyle \int_{\gamma_m} [v]\mu \, d\sigma = 0, \ \mu \in M_{h_m}(\gamma_m), \ 1 \le m \le M \big\} \ .$$

We remark that formally (1.13) is exactly the same definition as in (1.2). For simplicity, we use the same notations as before, but we point out that now V_h, u_h, and λ_h do not depend only on the order of the discretization and the triangulation but also on the special choice of M_h.

We briefly recall the two different mortar settings. The symmetric positive definite one can be written as: Find $u_h \in V_h$ such that

$$a(u_h, v_h) = (f, v_h)_0, \quad v_h \in V_h .\tag{1.14}$$

In the saddle point approach, the unconstrained product space X_h is independent of the choice of M_h. The abstract saddle point formulation has exactly the same structure as in (1.9). The only difference is that now the special Lagrange multiplier space (1.2) is replaced by the general one: Find $(u_h, \lambda_h) \in (X_h, M_h)$ such that

$$\begin{aligned} a(u_h, v_h) + b(\lambda_h, v_h) &= (f, v_h)_0, & v_h \in X_h , \\ b(\mu_h, u_h) \qquad\qquad &= 0, & \mu_h \in M_h . \end{aligned}\tag{1.15}$$

Within the general saddle point framework, the approximation property of V_h, is a consequence of the approximation property of X_h, the continuity of the bilinear form $b(\cdot, \cdot)$, and an inf-sup condition; see, e.g., [BF91]. A discrete inf-sup condition is necessary in order to obtain a priori estimates for the Lagrange multiplier. In the mortar situation, the bilinear form $b(\cdot, \cdot)$ is not uniformly continuous on $X_h \times M_h$; see (1.12). Thus one has to be very careful in the a priori analysis.

In the following three subsections, we show that we have the same quality of a priori estimates for the solutions u_h and (u_h, λ_h) as before for the standard case. Subsection 1.2.1 is devoted to the analysis of the best approximation property of the constrained space. It is based on the stability estimate in the $H_{00}^{1/2}$-norm for the generalized mortar projection. To establish this stability, assumptions (Sa), (Sc), and (Sd) are required.

In Subsect. 1.2.2, we consider the consistency error of the abstract nonconforming mortar formulation (1.14). In order to obtain a consistency error of at least the same order as the best approximation error, assumptions (Sa)–(Sd) are necessary. These are the two basic tools to obtain a priori estimates for $u - u_h$. To obtain the required order of the consistency error, the assumption (Sb) is of crucial importance. Roughly speaking, we have to require that the space $M_{h_m}(\gamma_m)$ contains the space of polynomials of order $\leq n_m - 1$.

The stability of the saddle point problem relies on a suitable inf-sup condition. We establish a discrete inf-sup conditions in Subsect. 1.2.3. Based on these preliminary considerations, we obtain a priori estimates for the Lagrange multiplier which are of the same order as the a priori estimates for $u - u_h$ in the energy norm.

In Subsect. 1.2.4, we present and analyze several examples for Lagrange multiplier spaces with particular emphasis on dual basis spaces. The advantage of those spaces satisfying a suitable biorthogonality relation is that the nodal basis functions of V_h have a local support. Examples are given in 2D and 3D and for piecewise quadratic finite elements. For higher order elements, we refer to [OW00].

1.2.1 An Approximation Property

The essential tool in the proof of the approximation property of the constrained space V_h is the stability of the mortar projection Π_m. In this subsection, we introduce a modified mortar projection which depends on the choice of the spaces $\widetilde{W}_{0;h_m}(\gamma_m)$ and $M_{h_m}(\gamma_m)$. The modified mortar projection, $\Pi_m : L^2(\gamma_m) \longrightarrow \widetilde{W}_{0;h_m}(\gamma_m)$, is defined by

$$\int_{\gamma_m} \Pi_m v \,\mu \, d\sigma = \int_{\gamma_m} v\mu \, d\sigma, \quad \mu \in M_{h_m}(\gamma_m) \ . \tag{1.16}$$

We observe that the dimension of $\widetilde{W}_{0;h_m}(\gamma_m)$ and $M_{h_m}(\gamma_m)$ is the same by construction. However, the mortar projection Π_m will not be well defined for arbitrary choices of $\widetilde{W}_{0;h_m}(\gamma_m)$ and $M_{h_m}(\gamma_m)$.

Remark 1.2. *In the original papers about mortar methods, the mortar projection for the standard Lagrange multiplier space is given in a different way; see Sect. 1.1. It satisfies an approximation property for H^1-functions, and can be extended to $H_{00}^{1/2}$-functions in a stable way. Here, we use a different form which does not take the values at the endpoints into account. However, both definitions give rise to operators which are identical when restricted to $H_{00}^{1/2}(\gamma_m)$. We note the following difference: The operator defined by (1.16) is in contrast to the one given by (1.8) not $H^1(\gamma_m)$-stable but $L^2(\gamma_m)$-stable. Figure 1.7 illustrates the stability properties of the two mortar projections.*

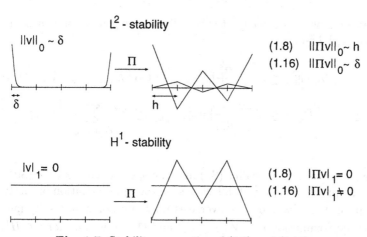

Fig. 1.7. Stability properties of (1.8) and (1.16)

In the following, we work with the mortar projection Π_m defined by (1.16). The stability of the mortar projection plays an essential role in the analysis of the best approximation error. The following lemma provides uniform stability in the L^2- and H^1-norms.

Lemma 1.3. *Under the assumptions (Sa), (Sc), and (Sd), the mortar projection (1.16) is well defined and is L^2-stable and H_0^1-stable:*

$$\|\Pi_m v\|_{0;\gamma_m} \leq C\|v\|_{0;\gamma_m}, \quad v \in L^2(\gamma_m) \ ,$$

$$|\Pi_m v|_{1;\gamma_m} \leq C|v|_{1;\gamma_m}, \quad v \in H_0^1(\gamma_m) \ .$$

Proof. The spectral equivalence (Sd) shows that M_1 is non-singular and thus that Π_m is well defined. Using the explicit representation $\Pi_m v = \sum_{i=1}^{N_m} \alpha_i \theta_i$, we find

$$M_1 \alpha = r$$

where $r \in \mathbb{R}^{N_m}$, with $r_i := \int_{\gamma_m} v\psi_i \, d\sigma$, $1 \leq i \leq N_m$. Due to the locality of the simply connected support of θ_i, $1 \leq i \leq N_m$, and the linear independence, we have

$$\|\Pi_m v\|_{0;\gamma_m}^2 \leq C \sum_{i=1}^{N_m} \alpha_i^2 \|\theta_i\|_{0;\gamma_m}^2 = C\alpha^T D_1 \alpha = Cr^T M_1^{-T} D_1 M_1^{-1} r \ .$$

Finally, the assumptions (Sa) and (Sd) yield the L^2-stability

$$\|\Pi_m v\|_{0;\gamma_m}^2 \leq Cr^T D_\psi^{-1} r = C \sum_{i=1}^{N_m} r_i^2 \|\psi_i\|_{0;\gamma_m}^{-2}$$

$$\leq C \sum_{i=1}^{N_m} \|v\|_{0;\mathrm{supp}\psi_i}^2 \leq C\|v\|_{0;\gamma_m}^2 \ .$$

Reasoning as before, we get the stability of the general mortar projection in a weighted L^2-norm

$$\sum_{e \in \mathcal{S}_{m;h_m}} \frac{1}{h_e^2} \|\Pi_m v\|_{0;e}^2 \leq C \sum_{i=1}^{N_m} \frac{\alpha_i^2}{h_{\theta_i}^2} \|\theta_i\|_{0;\gamma_m}^2 = C\alpha^T \widetilde{D}_1 \alpha = Cr^T M_1^{-T} \widetilde{D}_1 M_1^{-1} r$$

$$\leq Cr^T \widetilde{D}_\psi^{-1} r \leq C \sum_{i=1}^{N_m} \frac{1}{h_{\psi_i}^2} \|v\|_{0;\mathrm{supp}\psi_i}^2 \leq C \sum_{e \in \mathcal{S}_{m;h_m}} \frac{1}{h_e^2} \|v\|_{0;e}^2 \ .$$

$$(1.17)$$

Here, we have used the locality (Sa) and the spectral equivalence (Sd).

For the proof of the H_0^1-stability, we use the approximation property (Sc) of the space $\widetilde{W}_{0;h_m}(\gamma_m)$ and an inverse estimate for polynomials. By means of the best approximation $v_\theta \in \widetilde{W}_{0;h_m}(\gamma_m)$ and (1.17), we find for $v \in H_0^1(\gamma_m)$

$$|\Pi_m v|_{1;\gamma_m} \leq |\Pi_m(v - v_\theta)|_{1;\gamma_m} + |v_\theta|_{1;\gamma_m}$$

$$\leq C \left(\sum_{e \in \mathcal{S}_{m;h_m}} \frac{1}{h_e^2} \|\Pi_m(v - v_\theta)\|_{0;e}^2 \right)^{\frac{1}{2}} + |v_\theta|_{1;\gamma_m} \leq C|v|_{1;\gamma_m} \ . \ \square$$

The $H_{00}^{1/2}$-stability of the mortar projection is guaranteed by the L^2- and H_0^1-stability and an interpolation argument.

Following the lines of [BMP93, BMP94], it is easy to establish an approximation property for V_h in the 2D case provided that the mortar projection Π_m is $H_{00}^{1/2}$-stable. For each subdomain Ω_k, we use the Lagrange interpolation operator I_k, and define $w_h \in X_h$ by $w_{h|\Omega_k} := I_k u$. We note that w_h, in general, will not be contained in V_h. To obtain an element in V_h, we have to add appropriate corrections. We observe that the jump $[w_h]$ is in $H_0^1(\gamma_m)$ for each interface γ_m in 2D. This is not true for 3D, and the proof of the approximation property has to be modified. In the 2D case, we apply the mortar projection to $[w_h]$. The result $\Pi_m[w_h]$ is extended by zero onto $\partial\Omega_{n(m)} \setminus \gamma_m$ and the extension is still denoted by $\Pi_m[w_h]$. Then, the definition of $\widetilde{W}_{0;h_m}(\gamma_m)$ yields $\Pi_m[w_h] \in H^{1/2}(\partial\Omega_{n(m)})$. Now, $\Pi_m[w_h]$ is extended as a discrete harmonic function into the interior of $\Omega_{n(m)}$. Finally, we define

$$v_h := w_h - \sum_{m=1}^{M} \mathcal{H}_{n(m)}(\Pi_m[w_h]) , \qquad (1.18)$$

where $\mathcal{H}_{n(m)}$ denotes the discrete harmonic extension operator satisfying

$$\|\mathcal{H}_{n(m)}v\|_{1;\Omega_{n(m)}} \leq C\|v\|_{H^{\frac{1}{2}}(\partial\Omega_{n(m)})}, \quad v \in H^{\frac{1}{2}}(\partial\Omega_{n(m)}) ;$$

see [Bra66, Wid88]. Here, the extension $\mathcal{H}_{n(m)}(\Pi_m[w_h])$ vanishes outside $\Omega_{n(m)}$. By construction, we have

$$\int_{\gamma_m} [v_h]\mu \, d\sigma = \int_{\gamma_m} ([w_h] - \Pi_m[w_h])\mu \, d\sigma = 0, \quad \mu \in M_{h_m}(\gamma_m) ,$$

and thus $v_h \in V_h$. Replacing the test space M_h in the definition of the nonconforming space, we find that the approximation property is preserved as long as the modified mortar projection is $H_{00}^{1/2}$-stable.

Lemma 1.4. *Under the assumptions (Sa), (Sc) and (Sd) on $M_{h_m}(\gamma_m)$, $1 \leq m \leq M$, the nonconforming space V_h satisfies the approximation property in 2D,*

$$\inf_{v_h \in V_h} \|u - v_h\|_1^2 \leq C \sum_{k=1}^{K} h_k^{2n_k} |u|_{n_k+1;\Omega_k}^2 , \qquad (1.19)$$

if u is regular enough.

Proof. We set $v_h \in V_h$ as in (1.18) and find by the $H_{00}^{1/2}$-stability of Π_m and a coloring argument

$$\|u - v_h\|_1^2 \leq C \left(\|u - w_h\|_1^2 + \sum_{m=1}^M \|\Pi_m[w_h]\|_{H^{\frac{1}{2}}(\partial\Omega_{n(m)})}^2 \right)$$
$$\leq C \left(\|u - w_h\|_1^2 + \sum_{m=1}^M \|[u - w_h]\|_{H^{\frac{1}{2}}_{00}(\gamma_m)}^2 \right)$$
$$\leq C \sum_{k=1}^K h_k^{2n_k} |u|_{n_k+1;\Omega_k}^2 .$$

$\square$

We remark that the constant depends on the shape regularity of the triangulations and the decomposition but not on the number of subdomains. In particular, the ratio of the length of adjacent edges on the non-mortar sides enters in the bounds. But the ratio between the meshsizes on mortar and non-mortar sides does not enter. For a more detailed analysis of the constants in a special case, we refer to [Woh00a].

The 3D case. In the 3D case, we have to face the fact that the boundary of $\partial\gamma_m$ is a closed one dimensional curve. For the analysis of the standard mortar situation, we refer to [BM97, BD98, LSV94]. We cannot expect that $[w_h]|_{\gamma_m} = 0$ on $\partial\gamma_m$ in the case of non-matching meshes on $\partial\gamma_m$, where w_h is defined as before. Since the mortar projection given by (1.16) is L^2-stable but not H^1-stable, no uniform stability results are available for the $H^{1/2}_{00}$-norm. Thus, we cannot work with the discrete harmonic extension $\mathcal{H}_{n(m)}$ from $H^{1/2}(\partial\Omega_{n(m)})$ onto $H^1(\Omega_{n(m)})$ as in the 2D case. We replace $\mathcal{H}_{n(m)}$ by $\tilde{\mathcal{H}}_{n(m)}$ supported only in a small strip in the neighborhood of γ_m on the non-mortar side. It is defined by its values at the Lagrangian interpolation points x_p

$$\left(\tilde{\mathcal{H}}_{n(m)}v\right)(x_p) := \begin{cases} v(x_p), & x_p \in \gamma_m , \\ 0, & \text{elsewhere} . \end{cases}$$

Now, an inverse inequality yields, for $v \in \prod_{m=1}^M W_{h_m}(\gamma_m)$,

$$\sum_{m=1}^M \|\tilde{\mathcal{H}}_{n(m)}v\|_{1;\Omega_{n(m)}}^2 \leq C \sum_{m=1}^M \sum_{e \in \mathcal{S}_{m;h_m}} \frac{1}{h_e} \|v\|_{0;e}^2 .$$

Proceeding as in the 2D case and observing that Π_m is stable in this weighted L^2-norm finally yield an a priori estimate which depends on $(1 + h_{\text{mor}}/h_{\text{non}})$

$$\inf_{v_h \in V_h} \|u - v_h\|_1^2 \leq C \left(\sum_{k=1}^K h_k^{2n_k} |u|_{n_k+1;\Omega_k}^2 + \frac{h_{\text{mor}}}{h_{\text{non}}} \sum_{m=1}^M h_{\bar{n}(m)}^{2n_{\bar{n}(m)}} |u|_{n_{\bar{n}(m)}+1;\mathcal{S}_m}^2 \right) ,$$

where $h_{\text{mor}}/h_{\text{non}}$ stands for the maximum of the local ratio between meshsizes on the mortar and non-mortar sides, and $\mathcal{S}_m \subset \Omega_{\bar{n}(m)}$ is a strip of width $h_{\bar{n}(m)}$ and area $|\gamma_m|$ on the mortar side. We note that $|\mathcal{S}_m|/|\Omega_{\bar{n}(m)}|$ tends to zero if the meshsize tends to zero.

Although the constant in the a priori estimate depends on $(1 + h_{\text{mor}}/h_{\text{non}})$, it might be advantageous to associate the non-mortar side with the finer mesh.

We refer to Sect. 1.5.3, for numerical results illustrating the influence of the choice of the mortar method. In general, using very different meshsizes on the mortar and non-mortar sides is only appropriate if the coefficients are strongly discontinuous. Then, this jump of the coefficients is also reflected in the constants of the a priori estimates and this effect might cancel the factor $h_{\mathrm{mor}}/h_{\mathrm{non}}$. For a more detailed analysis of the constant in terms of the coefficients, we refer to [Woh00a]. In particular, we find that the ratio $a_{\mathrm{non}}/a_{\mathrm{mor}}$ enters into the constants of the a priori estimate in the energy norm. Here, a_{non} and a_{mor} stand for un upper and lower bound for the eigenvalues of the coefficient a on the non-mortar and mortar sides, respectively.

Remark 1.5. *We note that in the special situation that the triangulations $\mathcal{T}_{n(m);h_{n(m)}}$ and $\mathcal{T}_{\bar{n}(m);h_{\bar{n}(m)}}$ coincide on $\partial \gamma_m$; see Fig. 1.8, and $n_{n(m)} = n_{\bar{n}(m)}$, $1 \leq m \leq M$, we can proceed as in the 2D case. Then, applying the Lagrange interpolation yields $[w_h]|_{\gamma_m} = (I_{n(m)}u - I_{\bar{n}(m)}u)|_{\gamma_m} \in H_{00}^{1/2}(\gamma_m)$. We refer also to [KLPV00] for the special case $n_k = 1$, $1 \leq k \leq K$.*

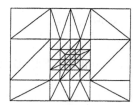 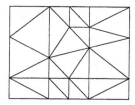

Fig. 1.8. Triangulations on mortar and non-mortar side of γ_m in 3D

1.2.2 The Consistency Error

The definition of V_h guarantees that the jump of an element is orthogonal to the Lagrange multiplier space M_h. However, in general, the L^2-norm of the jump will not vanish. Thus, we are in a nonconforming setting and $V_h \not\subset H_0^1(\Omega)$. In this case, the approximation property (1.19) is not sufficient to obtain optimal a priori estimates for the finite element solution of (1.14). According to Strang's Lemma, the consistency error

$$\sup_{v_h \in V_h} \frac{\sum\limits_{m=1}^{M} \int\limits_{\gamma_m} a \frac{\partial u}{\partial \mathbf{n}} [v_h] \, d\sigma}{\|v_h\|_1}$$

has to be considered; see, e.g., [Bra97].

The weighted L^2-norm of the jump of an element in V_h measures its nonconformity. Using the definition of the mortar projection, we find that for an element $v \in V_h$,

$$\Pi_m(v_{|\Omega_{\bar{n}(m)}}) = \Pi_m(v_{|\Omega_{n(m)}}), \quad 1 \le m \le M , \tag{1.20}$$

where $v_{|\Omega_{\bar{n}(m)}}$ is the trace of v on the mortar side and $v_{|\Omega_{n(m)}}$ is the trace of v on the non-mortar side. However, Π_m satisfies an approximation property only in $H_0^1(\gamma_m)$ but not in $H^1(\gamma_m)$. Since no continuity is imposed on the space V_h at the crosspoints, the jump $[v]_{|\gamma_m}$, $v \in V_h$, will be in general not belong to $H_0^1(\gamma_m)$. In the first mortar papers, the continuity at the crosspoints was required and both properties, the approximation property on $H^1(\gamma_m)$ and (1.20), could be obtained at the same time for the mortar projection. For the proof of the approximation property of the constrained space in 2D, it was important to work with an operator guaranteeing that the result is zero on $\partial \gamma_m$ and can be extended by zero onto $\partial \Omega_{n(m)} \setminus \gamma_m$ resulting in an element in $H^{1/2}(\partial \Omega_{n(m)})$.

Here, we introduce a new projection having both properties even without continuity at the crosspoints, but which does not guarantee that the result is zero on $\partial \gamma_m$. In the definition of the mortar projection, we replace $\widetilde{W}_{0;h_m}(\gamma_m)$ by $\widetilde{W}_{h_m}(\gamma_m)$ and define $P_m : L^2(\gamma_m) \longrightarrow \widetilde{W}_{h_m}(\gamma_m)$ by

$$\int\limits_{\gamma_m} P_m v \, \mu \, d\sigma = \int\limits_{\gamma_m} v \mu \, d\sigma, \quad \mu \in M_{h_m}(\gamma_m) .$$

The projection P_m is well defined under the assumption (Sd). By definition, we find, for $v \in V_h$,

$$P_m[v] = 0, \quad 1 \le m \le M .$$

We remark that P_m cannot replace Π_m in the proof of the approximation property. For the construction of an element in V_h satisfying the approximation property, it is important that $H_{00}^{1/2}$-functions are mapped onto $H_{00}^{1/2}$-functions. This is the case for Π_m but not for P_m.

In addition to P_m, we now introduce a dual operator $Q_m : L^2(\gamma_m) \longrightarrow M_{h_m}(\gamma_m)$ by

$$\int\limits_{\gamma_m} Q_m v \, \theta \, d\sigma = \int\limits_{\gamma_m} v \theta \, d\sigma, \quad \theta \in \widetilde{W}_{h_m}(\gamma_m) .$$

The following lemma shows that P_m and Q_m satisfy certain approximation properties.

Lemma 1.6. *Under the assumptions (Sa), (Sc), and (Sd), there exist constants such that*

$$\sum_{e \in \mathcal{S}_{m;h_m}} \frac{1}{h_e} \|v - P_m v\|_{0;e}^2 \le C|v|_{\frac{1}{2};\gamma_m}^2, \qquad v \in H^{\frac{1}{2}}(\gamma_m) ,$$

$$\sum_{e \in \mathcal{S}_{m;h_m}} h_e \|v - Q_m v\|_{0;e}^2 \le C h_m^{2n_m} |v|_{n_m-\frac{1}{2};\gamma_m}^2, \qquad v \in H^{n_m-\frac{1}{2}}(\gamma_m) ,$$

$$\sum_{e \in \mathcal{S}_{m;h_m}} h_e \|v - Q_m v\|_{0;e}^2 \ge c \, \|v - Q_m v\|_{(H_{00}^{\frac{1}{2}}(\gamma_m))'}^2, \qquad v \in L^2(\gamma_m) .$$

Proof. We start with the proof of the approximation property of $P_m v$. It is an easy consequence of the locality of the supports, (Sa), the approximation property, (Sc), of $\widetilde{W}_{h_m}(\gamma_m)$, and the spectral equivalence, (Sd). As in the proof of Lemma 1.3, (Sa) and (Sd) guarantee the stability of P_m in the L^2- and a weighted L^2-norm. Finally, (Sc) yields for $s = 1/2$

$$\sum_{e \in \mathcal{S}_{m;h_m}} \frac{1}{h_e} \|v - P_m v\|_{0;e}^2 = \sum_{e \in \mathcal{S}_{m;h_m}} \frac{1}{h_e} \|v - v_\theta - P_m(v - v_\theta)\|_{0;e}^2$$

$$\leq C \sum_{e \in \mathcal{S}_{m;h_m}} \frac{1}{h_e} \|v - v_\theta\|_{0;e}^2 \leq C |v|_{\frac{1}{2};\gamma_m}^2 .$$

To prove the second inequality, we first consider the stability property of Q_m in a weighted L^2-norm. Observing that $\int_{\gamma_m} Q_m v \, w \, d\sigma = \int_{\gamma_m} v \, P_m w \, d\sigma$, $v, w \in L^2(\gamma_m)$ and using the stability of P_m in the weighted dual L^2-norm, we get

$$\sum_{e \in \mathcal{S}_{m;h_m}} h_e \|Q_m v\|_{0;e}^2 = \sup_{w \in L^2(\gamma_m)} \frac{\left(\int_{\gamma_m} v \, P_m w \, d\sigma \right)^2}{\sum_{e \in \mathcal{S}_{m;h_m}} \frac{1}{h_e} \|w\|_{0;e}^2}$$

$$\leq \sum_{e \in \mathcal{S}_{m;h_m}} h_e \|v\|_{0;e}^2 \sup_{w \in L^2(\gamma_m)} \frac{\sum_{e \in \mathcal{S}_{m;h_m}} \frac{1}{h_e} \|P_m w\|_{0;e}^2}{\sum_{e \in \mathcal{S}_{m;h_m}} \frac{1}{h_e} \|w\|_{0;e}^2} \leq C \sum_{e \in \mathcal{S}_{m;h_m}} h_e \|v\|_{0;e}^2 .$$

Now as before, the approximation property of Q_m follows from the stability and the best approximation property (Sb): Let $v_\psi \in M_{h_m}(\gamma_m)$ such that (Sb) holds. Then, we find

$$\sum_{e \in \mathcal{S}_{m;h_m}} h_e \|v - Q_m v\|_{0;e}^2 = \sum_{e \in \mathcal{S}_{m;h_m}} h_e \|v - v_\psi - Q_m(v - v_\psi)\|_{0;e}^2$$

$$\leq C \sum_{e \in \mathcal{S}_{m;h_m}} h_e \|v - v_\psi\|_{0;e}^2 \leq C h_m^{nm} |v|_{n_m - \frac{1}{2};\gamma_m}^2 .$$

The proof of the last inequality is mainly based on the fact that Q_m is the dual operator of P_m. Using the definition of the dual norm and the first inequality of Lemma 1.6, we obtain

$$\|v - Q_m v\|_{(H_{00}^{\frac{1}{2}}(\gamma_m))'} = \sup_{w \in H_{00}^{\frac{1}{2}}(\gamma_m)} \frac{\int_{\gamma_m} (v - Q_m v)(w - P_m w) \, d\sigma}{\|w\|_{H_{00}^{\frac{1}{2}}(\gamma_m)}}$$

$$\leq \left(\sum_{e \in \mathcal{S}_{m;h_m}} h_e \|v - Q_m v\|_{0;e}^2 \right)^{\frac{1}{2}} \sup_{w \in H_{00}^{\frac{1}{2}}(\gamma_m)} \frac{\left(\sum_{e \in \mathcal{S}_{m;h_m}} \frac{1}{h_e} \|w - P_m w\|_{0;e}^2 \right)^{\frac{1}{2}}}{\|w\|_{H_{00}^{\frac{1}{2}}(\gamma_m)}}$$

$$\leq C \left(\sum_{e \in \mathcal{S}_{m;h_m}} h_e \|v - Q_m v\|_{0;e}^2 \right)^{\frac{1}{2}} . \qquad \square$$

An upper bound for the weighted L^2-norm of the jump of an element in V_h can be obtained by means of the projection P_m. In particular, the nonconformity of an element can be measured by this norm. The following lemma provides an upper bound for the jump in the weighted L^2-norm.

Lemma 1.7. *Under the assumptions (Sa), (Sc), and (Sd), the weighted L^2-norm of the jumps of an element $v \in V_h$ is bounded by*

$$\sum_{m=1}^{M} \sum_{e \in \mathcal{S}_{m;h_m}} \frac{1}{h_e} \||[v]|\|_{0;e}^2 \leq \inf_{w \in H_0^1(\Omega)} \|v - w\|_1^2 \ . \tag{1.21}$$

Proof. The proof uses the same ideas as in the case of the standard constrained space; see [Woh99a]. Using the orthogonality of the jump and the trial space and Lemma 1.6, we find for $v \in V_h$

$$\sum_{e \in \mathcal{S}_{m;h_m}} \frac{1}{h_e} \||[v]|\|_{0;e}^2 = \sum_{e \in \mathcal{S}_{m;h_m}} \frac{1}{h_e} \||[v] - P_m[v]|\|_{0;e}^2 \leq C \||[v]|\|_{\frac{1}{2};\gamma_m} \ .$$

Finally, using the continuity of the trace operator, we get for each $w \in H_0^1(\Omega)$,

$$\sum_{e \in \mathcal{S}_{m;h_m}} \frac{1}{h_e} \||[v]|\|_{0;e}^2 \leq C \left(|v|_{\Omega_{n(m)}} - w|_{\frac{1}{2};\gamma_m}^2 + |v|_{\Omega_{\bar{n}(m)}} - w|_{\frac{1}{2};\gamma_m}^2 \right)$$

$$\leq C \left(\|v - w\|_{1;\Omega_{n(m)}}^2 + \|v - w\|_{1;\Omega_{\bar{n}(m)}}^2 \right) \ .$$

Summing over the interfaces γ_m gives (1.21) with a constant independent of the number of subdomains. □

The consistency error of the mortar formulation (1.14) is closely related to the nonconformity of the elements in the constrained space V_h. The following lemma provides an upper bound for the consistency error. The proof is based on Lemma 1.7 and the approximation property (Sb) of M_h.

Lemma 1.8. *Under the assumptions (Sa)–(Sd) and u regular enough, there exists a constant such that*

$$\sup_{v_h \in V_h} \frac{\sum_{m=1}^{M} \int_{\gamma_m} a \frac{\partial u}{\partial n} [v_h] \, d\sigma}{\|v_h\|_1} \leq C \left(\sum_{k=1}^{K} h_k^{2n_k} |u|_{n_k+1;\Omega_k}^2 \right)^{\frac{1}{2}} \ .$$

Proof. By means of the orthogonality, we find for $v_h \in V_h$ and $\mu_h \in M_{h_m}(\gamma_m)$ that

$$\int_{\gamma_m} a \frac{\partial u}{\partial n} [v_h] \, d\sigma = \int_{\gamma_m} \left(a \frac{\partial u}{\partial n} - \mu_h \right) [v_h] \, d\sigma$$

$$\leq C \left(\sum_{e \in \mathcal{S}_{m;h_m}} h_e \| a \frac{\partial u}{\partial n} - \mu_h \|_{0;e}^2 \right)^{\frac{1}{2}} \left(\sum_{e \in \mathcal{S}_{m;h_m}} \frac{1}{h_e} \||[v_h]|\|_{0;e}^2 \right)^{\frac{1}{2}} \ .$$

Using $w = 0$ in Lemma 1.7, the approximation property (Sb), a trace theorem and summing over the interfaces guarantee a consistency error of optimal order.   □

The coercivity of the bilinear form $a(\cdot, \cdot)$ on $V_h \times V_h$ is an easy consequence of (Sb). From (Sb), we obtain $P_0(\gamma_m) \subset M_{h_m}(\gamma_m)$ which yields $V_h \subset Y$. Thus, the unique solvability of (1.14) is guaranteed. Now, the approximation property of V_h and the consistency error guarantee an optimal order discretization scheme. We obtain the same quality of the a priori estimates as in the standard case (1.4). The stability of the mortar projection yields the approximation property. This argument is mainly based on the spectral equivalence (Sd) and the approximation property of $\widetilde{W}_{h_m}(\gamma_m)$. The essential tool in the proof of the consistency error is the approximation property of $M_{h_m}(\gamma_m)$. Using Lemmas 1.4 and 1.8, we obtain standard a priori estimates for the modified mortar approach (1.14) in the broken H^1-norm if the solution u is smooth enough, i.e.,

$$\|u - u_h\|_1^2 \leq C \sum_{k=1}^{K} h_k^{2 n_k} |u|_{n_k+1;\Omega_k}^2 .\tag{1.22}$$

If we furthermore assume H^2-regularity, the discretization error $u - u_h$ in the L^2-norm is of order h^2. The proof is based on the Aubin–Nitsche trick. For the standard mortar setting it can be found, e.g., in [BDW99]. Introducing the dual problems: Find $w \in H_0^1(\Omega)$ and $w_h \in V_h$ such that

$$a(w, v) = (u - u_h, v)_0, \quad v \in H_0^1(\Omega), \quad a(w_h, v) = (u - u_h, v)_0, \quad v \in V_h ,$$

we get

$$\|u - u_h\|_0^2 = a(w - w_h, u - u_h) - \sum_{m=1}^{M} \left(\int_{\gamma_m} a \frac{\partial w}{\partial \mathbf{n}} [u_h] \, d\sigma - \int_{\gamma_m} a \frac{\partial u}{\partial \mathbf{n}} [w_h] \, d\sigma \right) .$$

Then, the H^2-regularity, Lemma 1.7, (Sb), and observing that the jump of an element in V_h is orthogonal on M_h yield

$$\|u - u_h\|_0^2 \leq a(u - u_h, u - u_h)^{1/2} a(w - w_h, w - w_h)^{1/2}$$
$$+ \left(\sum_{m=1}^{M} \sum_{e \in \mathcal{S}_{m;h_m}} \frac{1}{h_e} \|[w_h]\|_{0;e}^2 \right)^{1/2} \inf_{\mu_h \in M_h} \left(\sum_{m=1}^{M} \sum_{e \in \mathcal{S}_{m;h_m}} h_e \|\lambda - \mu_h\|_{0;e}^2 \right)^{1/2}$$
$$+ \left(\sum_{m=1}^{M} \sum_{e \in \mathcal{S}_{m;h_m}} \frac{1}{h_e} \|[u_h]\|_{0;e}^2 \right)^{1/2} \inf_{\mu_h \in M_h} \left(\sum_{m=1}^{M} \sum_{e \in \mathcal{S}_{m;h_m}} h_e \|\chi - \mu_h\|_{0;e}^2 \right)^{1/2}$$
$$\leq C \left(\|u - u_h\|_1 (\|w - w_h\|_1 + h\|w\|_2) + \|w - w_h\|_1 \inf_{\mu_h \in M_h} \|\lambda - \mu_h\|_{-\frac{1}{2};S} \right)$$

where $\chi|_{\gamma_m} := a \frac{\partial w}{\partial \mathbf{n}}$. Using the a priori estimate for the energy norm (1.22) and the approximation property (Sb), we obtain an order h^2 a priori estimate for the L^2-norm

$$\|u - u_h\|_0 \leq Ch \left(\|u - u_h\|_1 + \inf_{\mu_h \in M_h} \|\lambda - \mu_h\|_{h - \frac{1}{2};\mathcal{S}} \right)$$

$$\leq Ch \left(\sum_{k=1}^{K} h_k^{2n_k} |u|_{n_k+1;\Omega_k}^2 \right)^{\frac{1}{2}} . \tag{1.23}$$

We remark that the global meshsize h enters into the a priori estimates in the L^2-norm and not only those of the subdomains. This is in contrast with the H^1-norm estimate.

1.2.3 Discrete Inf-sup Conditions

The approximation property of M_h is in general not sufficient to obtain a priori estimates for the Lagrange multiplier. In addition, suitable inf-sup conditions have to be satisfied; see, e.g., [BF91]. In this subsection, we analyze the error in the Lagrange multiplier for the $H_{00}^{1/2}$-dual norm and a weighted L^2-norm. For both these norms, we obtain a priori estimates of the same quality as for the standard mortar approach. In the following, we show that the inf-sup conditions, established in [Ben99] and [Woh99a] for the standard case, also holds for the general pairing (X_h, M_h).

Before we prove the discrete inf-sup conditions, we consider an extension operator $\mathcal{H}_{M_m} : \widetilde{W}_{0;h_m}(\gamma_m) \longrightarrow X_{h_{n(m)};n_{(m)}}$. Its definition depends on the choice of $\|\cdot\|_{M'}$. In Sect. 1.1, the norm, $\|\cdot\|_{M'}$, associated with the Lagrange multiplier space was defined in two different ways. For both definitions, we consider a dual norm.

We start with the case $\|\cdot\|_{M_m'} = \|\cdot\|_{(H_{00}^{1/2}(\gamma_m))'}$ and set

$$\|v\|_{M_m} := \|v\|_{H_{00}^{\frac{1}{2}}(\gamma_m)}, \quad v \in M_m := H_{00}^{\frac{1}{2}}(\gamma_m) .$$

The extension operator $\mathcal{H}_{M_m}$ is then defined as the discrete harmonic extension, where the elements of $\widetilde{W}_{0;h_m}(\gamma_m)$ are extended by zero onto $\partial\Omega_{n(m)} \setminus \gamma_m$.

In the case of $\|\cdot\|_{M_m'} = \|\cdot\|_{h_m^{-1/2};\gamma_m}$, we set $M_m := L^2(\gamma_m)$ and $\|\cdot\|_{M_m} := \|\cdot\|_{h_m^{1/2};\gamma_m}$. The mesh dependent norm $\|\cdot\|_{h_m^{1/2};\gamma_m}$ has the inverse weight compared with $\|\cdot\|_{h_m^{-1/2};\gamma_m}$

$$\|\mu\|_{h_m^{\frac{1}{2}};\gamma_m}^2 := \sum_{e \in \mathcal{S}_{m;h_m}} \frac{1}{h_e} \|\mu\|_{0;e}^2, \quad \mu \in L^2(\gamma_m) .$$

The corresponding operator $\mathcal{H}_{M_m}$ is defined as the trivial extension by zero, i.e., we set all nodal values on $\partial\Omega_{n(m)} \setminus \gamma_m$ and in $\Omega_{n(m)}$ to zero. Then, $\mathcal{H}_{M_m}\phi$ is non zero only in a strip of area $|\gamma_m|$ and of width h_m.

For both cases, we obtain the following stability property

$$\|\mathcal{H}_{M_m}\phi\|_{1;\Omega_{n(m)}} \leq C\|\phi\|_{M_m}, \quad \phi \in \widetilde{W}_{0;h_m}(\gamma_m) .$$

Based on this stability estimate, we can easily establish the following discrete inf-sup condition for both choices of $\|\cdot\|_{M'}$.

Lemma 1.9. *Under the assumptions (Sa), (Sc) and (Sd), there exists a constant independent of h such that*

$$\inf_{\substack{\mu_h \in M_h \\ \mu_h \neq 0}} \sup_{\substack{v_h \in X_h \\ v_h \neq 0}} \frac{b(v_h, \mu_h)}{\|\mu_h\|_{M'} \|v_h\|_1} \geq c \ . \tag{1.24}$$

Proof. Considering one interface γ_m at a time, we get, by means of the stability of the mortar projection in the M_m-norm,

$$\|\mu_h\|_{M'_m} = \sup_{\phi \in M_m} \frac{(\mu_h, \phi)_{0;\gamma_m}}{\|\phi\|_{M_m}} = \sup_{\phi \in M_m} \frac{(\mu_h, \Pi_m\phi)_{0;\gamma_m}}{\|\phi\|_{M_m}}$$

$$\leq C \sup_{\phi \in M_m} \frac{(\mu_h, \Pi_m\phi)_{0;\gamma_m}}{\|\Pi_m\phi\|_{M_m}} \leq C \max_{\phi \in \widetilde{W}_{0;h_m}(\gamma_m)} \frac{(\mu_h, \phi)_{0;\gamma_m}}{\|\phi\|_{M_m}} \ .$$

The maximizing element in $\widetilde{W}_{0;h_m}(\gamma_m)$ with M_m-norm equal one, is called ϕ_{μ_h}. To obtain an element $v_m \in X_h$ from $\phi_{\mu_h} \in \widetilde{W}_{0;h_m}(\gamma_m)$, we apply the extension operator $\mathcal{H}_{M_m}$

$$v_m|_{\Omega \setminus \Omega_{n(m)}} := 0, \qquad v_m|_{\overline{\Omega}_{n(m)}} := \mathcal{H}_{M_m}\phi_{\mu_h} \ .$$

Now, we find that for both choices for $\| \cdot \|_{M'_m}$

$$0 \leq (\mu_h, \phi_{\mu_h})_{0;\gamma_m} = b(v_m, \mu_h) \ ,$$

and that $\|v_m\|_1 \leq C|v_m|_{1;\Omega_{n(m)}} \leq C\|\phi_{\mu_h}\|_{M_m} = C$. Finally, we set

$$v_{\mu_h} := \sum_{m=1}^{M} b(v_m, \mu_h)v_m \ ,$$

and observe that $v_{\mu_h} = 0$ if and only if $\mu_h = 0$. Introducing $\mathcal{N}(k) := \{1 \leq m \leq M \mid n(m) = k\}$ and observing that the number of elements in $\mathcal{N}(k)$ is bounded independently of the number of subdomains, we find

$$\|v_{\mu_h}\|_1^2 = \| \sum_{k=1}^{K} \sum_{m \in \mathcal{N}(k)} b(v_m, \mu_h)v_m\|_1^2 = \sum_{k=1}^{K} \| \sum_{m \in \mathcal{N}(k)} b(v_m, \mu_h)v_m\|_{1;\Omega_k}^2$$

$$\leq C \sum_{k=1}^{K} \sum_{m \in \mathcal{N}(k)} \|b(v_m, \mu_h)v_m\|_{1;\Omega_k}^2 \leq C \sum_{m=1}^{M} b(v_m, \mu_h)^2$$

$$\leq C \sum_{m=1}^{M} \|\mu_h\|_{M'_m}^2 \|\phi_{\mu_h}\|_{M_m}^2 = C \sum_{m=1}^{M} \|\mu_h\|_{M'_m}^2 = C \|\mu_h\|_{M'}^2 \ .$$

Summing over all interfaces yields

$$\|\mu_h\|_{M'}^2 \leq C \sum_{m=1}^{M} b(v_m, \mu_h)^2 = C b(v_{\mu_h}, \mu_h) \leq C \frac{b(v_{\mu_h}, \mu_h)}{\|v_{\mu_h}\|_1} \|\mu_h\|_{M'} \ .$$

By construction, we have found for each $\mu_h \in M_h$, $\mu_h \neq 0$, a non zero $v_{\mu_h} \in X_h$ such that

$$\|\mu_h\|_{M'} \leq C \, \frac{b(v_{\mu_h}, \mu_h)}{\|v_{\mu_h}\|_1} \, .$$

$\square$

We remark that by construction, the inf-sup condition (1.24) also holds true if the supremum over X_h is replaced by the supremum over a suitable subspace. Moreover, two stability estimates hold. The first one is obtained by a coloring argument and is already used in the proof of (1.24)

$$\|v_{\mu_h}\|_1 \leq C \, \|\mu_h\|_{M'} \, . \tag{1.25}$$

The second follows from the definition of v_{μ_h} and the duality of the norms

$$\begin{aligned}
\|[v_{\mu_h}]\|_M^2 &= \sum_{m=1}^{M} (b(v_m, \mu_h))^2 \|[v_m]\|_{M_m}^2 = \sum_{m=1}^{M} (b(v_m, \mu_h))^2 \\
&\leq \sum_{m=1}^{M} \|\mu_h\|_{M'_m}^2 \|[v_m]\|_{M_m}^2 = \|\mu_h\|_{M'}^2 \, .
\end{aligned} \tag{1.26}$$

The inf-sup condition (1.24) together with the approximation property, Lemma 1.6, and the first equation of the saddle point problem give an a priori estimate similar to (1.22) for the Lagrange multiplier.

Lemma 1.10. *Under the assumptions (Sa)–(Sd) and u regular enough, the following a priori estimate for the Lagrange multiplier holds*

$$\|\lambda - \lambda_h\|_{M'}^2 \leq C \sum_{k=1}^{K} h_k^{2n_k} |u|_{n_k+1;\Omega_k}^2 \, . \tag{1.27}$$

Proof. Following [Ben99] and using the first equation of the saddle point problem, we get

$$b(v_h, \mu_h - \lambda_h) = a(u_h - u, v_h) + b(v_h, \mu_h - \lambda), \quad v_h \in X_h \, .$$

Choosing $v_{\mu_h - \lambda_h}$ as in the proof of (1.24), and applying the stability estimates (1.25) and (1.26), we find

$$\begin{aligned}
\|\mu_h - \lambda_h\|_{M'}^2 &\leq C \, b(v_{\mu_h - \lambda_h}, \mu_h - \lambda_h) \\
&= C \, (a(u_h - u, v_{\mu_h - \lambda_h}) + b(v_{\mu_h - \lambda_h}, \mu_h - \lambda)) \\
&\leq C \, (\|u - u_h\|_1 \|v_{\mu_h - \lambda_h}\|_1 + \|\mu_h - \lambda\|_{M'} \|[v_{\mu_h - \lambda_h}]\|_M) \\
&\leq C \, (\|u - u_h\|_1 + \|\mu_h - \lambda\|_{M'}) \|\mu_h - \lambda_h\|_{M'} \, .
\end{aligned}$$

Applying the triangle inequality and choosing $\mu_h|_{\gamma_m} := Q_m \lambda|_{\gamma_m}$, $1 \leq m \leq M$, we find by means of Lemma 1.6

$$\|\lambda - \lambda_h\|_{M'}^2 \leq C \left(\|u - u_h\|_1 + \sum_{m=1}^{M} h_m^{2n_m} |\lambda|_{n_m - \frac{1}{2}; \gamma_m}^2 \right)$$

$$\leq C \sum_{k=1}^{K} h_k^{2n_k} |u|_{n_k + 1; \Omega_k}^2 \; .$$

Here, we have used that λ restricted to γ_m is $a\nabla u \cdot \mathbf{n}$ and a trace theorem.
□

There is a structural difference between the 2D and 3D case only in the proof of the approximation property. The proofs of the consistency error and the discrete inf-sup condition are exactly the same.

Remark 1.11. *The a priori estimates (1.22), (1.23) and (1.27) can be weakened for the more general case* $u \in H^t(\Omega) \cap \prod_{k=1}^{K} H^{s_k}(\Omega_k)$, $s_k \geq t > 3/2$. *We refer to [Woh00a] for the lowest order case.*

1.2.4 Examples of Lagrange Multiplier Spaces

In the previous subsections, a general framework was given for new Lagrange multiplier spaces. Here, we focus on Lagrange multiplier spaces based on a dual basis and present concrete examples. The main advantage of such a biorthogonal basis is the locality of the supports of the nodal basis functions of the constrained space V_h. Additionally, the implementation of the mortar method for such a basis can be carried out using the unconstrained product space; details are provided in Sect. 1.3. We consider the special case where $\{\psi_i, 1 \leq i \leq N_m\}$ and $\{\theta_i, 1 \leq i \leq N_m\}$ are biorthogonal and satisfy

$$\text{(Se)} \quad \int_{\gamma_m} \theta_i \psi_j \, d\sigma = \delta_{ij} c_i \int_{\gamma_m} \theta_i^2 \, d\sigma, \quad c \leq c_i \leq C \; .$$

Considering now the support of an element $v_h \in V_h$, we find a structural difference between the standard case and the more general one satisfying (Sa)–(Se). We note that $\widehat{W}_{0;h_m}(\gamma_m) \subset W_{h_m}(\gamma_m)$, thus $\{\theta_i \,|\, 1 \leq i \leq N_m\}$ can be easily extended to a basis of $W_{h_m}(\gamma_m)$ with local support; $\{\theta_i \,|\, 1 \leq i \leq \nu_m\}$, $\nu_m := \dim W_{h_m}(\gamma_m)$. We recall that $N_m < \nu_m$. Furthermore in 2D, we find that $N_m \leq \nu_m - 2$. In the 3D case, we have $N_m \leq \nu_m - N_p$, where N_p is the number of nodes on $\partial \gamma_m$. Each $v \in X_h$, restricted to a non-mortar side γ_m, can be written as

$$v|_{\gamma_m} = \sum_{i=1}^{\nu_m} \alpha_i \theta_i \; . \tag{1.28}$$

Let $v \in X_h$ restricted to γ_m be given as in (1.28). Then, $v \in V_h$ if and only if for each non-mortar side γ_m

$$\alpha_i = \frac{\int_{\gamma_m} (v|_{\Omega_{\bar{n}(m)}} - \sum_{j=1+N_m}^{\nu_m} \alpha_j \theta_j) \psi_i \, d\sigma}{c_i \int_{\gamma_m} \theta_i^2 \, d\sigma}, \quad 1 \leq i \leq N_m \; . \tag{1.29}$$

The proof follows from (1.28) and the biorthogonality relation (Se). Using (1.29) and taking (Sa) into account, it is easy to construct nodal basis functions of V_h that have local support. As in the standard finite element context, nodal basis functions can be defined such that the diameter of the support is bounded by Ch. Here, C depends on the maximum number of edges on the mortar and non-mortar sides that have a non-empty intersection with the supports of ψ_i and θ_i, $1 \leq i \leq N_m$, respectively; see Fig. 1.9. We recall that this is not possible in the standard mortar case; comparing Fig. 1.5 with Fig. 1.9 shows the structural difference. In contrast to the standard case, the value of an element $v \in V_h$ at a point p on the non-mortar side is determined completely by its values in a small neighborhood of p on the mortar side. As a consequence, the constraints at the interfaces can be locally satisfied.

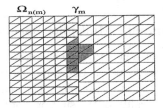

Fig. 1.9. Structure of the support of a nodal basis function in V_h, (dual)

Working with a biorthogonal basis reflects the duality between the trace space of the weak solution and the one of the flux. The basis functions of such a dual basis are, in general, not continuous and cannot be defined as a trace of conforming finite elements. The approximation property, well known for trace spaces, has, in general, to be checked by hand. The following lemma provides a simple tool for verifying the approximation property (Sb) of a Lagrange multiplier space satisfying the biorthogonality relation (Se).

Lemma 1.12. *Under the assumptions (Sa), (Sd), and (Se), the Lagrange multiplier space $M_{h_m}(\gamma_m)$ satisfies (Sb) if and only if $P_{n_m-1}(\gamma_m) \subset M_{h_m}(\gamma_m)$.*

Proof. The proof is based on arguments similar to the Bramble–Hilbert Lemma. Obviously, $P_{n_m-1}(\gamma_m) \subset M_{h_m}(\gamma_m)$ is a necessary condition for the approximation property (Sb). We define $\psi_v := \sum_{i=1}^{N_m} \alpha_i(v)\psi_i$ for $v \in H^{-1/2}(\gamma_m)$ by

$$\alpha_i(v) := \frac{\int\limits_{\gamma_m} v\theta_i \, d\sigma}{c_i \int\limits_{\gamma_m} \theta_i^2 \, d\sigma}, \quad 1 \leq i \leq N_m .$$

Then, the spectral equivalence (Sd) and the fact that M_1 is a diagonal matrix yield $\alpha_i^2(v)\|\psi_i\|_{0;\gamma_m}^2 \leq C\|v\|_{0; \, \mathrm{supp}\theta_i}^2$ for $v \in L^2(\gamma_m)$. Considering an element $e \in \mathcal{S}_{m;h_m}$ at a time and using the locality of the supports of θ_i and ψ_i, we obtain the following stability estimate

$$\|\psi_v\|^2_{0;e} \le C \sum_{\substack{i=1 \\ e \cap \mathrm{supp}\,\psi_i \ne \emptyset}}^{N_m} \alpha_i^2(v)\|\psi_i\|^2_{0;\gamma_m} \le C \sum_{\substack{i=1 \\ e \cap \mathrm{supp}\,\psi_i \ne \emptyset}}^{N_m} \|v\|^2_{0;\mathrm{supp}\,\theta_i} \le C\|v\|^2_{0;D_e} \ .$$

Here, $e \subset D_e$ is a simply connected union of at most a fixed number of elements $e' \in \mathcal{S}_{m;h_m}$. The maximum number of elements contained in D_e depends on the number of elements contained in the supports of θ_i and ψ_i but not on the meshsize. If $P_{n_m-1}(\gamma_m) \subset M_{h_m}(\gamma_m)$, then $\psi_v = v$ for $v \in P_{n_m-1}(\gamma_m)$, and we find

$$\begin{aligned}
\|v - \psi_v\|_{0;e} &\le \|v - \Pi_{n_m-1}v\|_{0;e} + \|\psi_{(v-\Pi_{n_m-1}v)}\|_{0;e} \\
&\le C\|v - \Pi_{n_m-1}v\|_{0;D_e} \le Ch_{D_e}^{n_m-\frac{1}{2}}|v|_{n_m-\frac{1}{2};D_e} \ ,
\end{aligned} \tag{1.30}$$

where h_{D_e} is the diameter of D_e, and Π_{n_m-1} is a locally defined L^2-projection onto $P_{n_m-1}(\gamma_m)$

$$\int_{D_e} \Pi_{n_m-1}v \ w \, d\sigma = \int_{D_e} v w \, d\sigma, \quad w \in P_{n_m-1}(\gamma_m) \ .$$

The local quasi-uniformity of the triangulation yields

$$ch_{e'} \le h_e \le Ch_{e'}, \quad e' \in D_e \ .$$

Together with the fact that the number of elements in D_e is bounded by a constant, we obtain (Sb) by summing over the elements and using (1.30).
□

The rest of this subsection is devoted to the construction of alternative Lagrange multiplier spaces, which provide optimal finite element solutions. We restrict ourselves to low order finite elements and refer to [OW00] for the general order case in 2D. As has been shown earlier, it is sufficient to verify (Sa)–(Sd). Furthermore if, in addition, the biorthogonality relation (Se) is satisfied, a basis of V_h having local support can be constructed. The idea of using dual spaces for the definition of the Lagrange multiplier space can also be carried over to 3D. As in the standard mortar approach [BM97, BD98], the analysis and the definition of the Lagrange multiplier space is more technical than in 2D.

1.2.4.1 The First Order Case in 2D. We start with the first order case, i.e., $n_k = 1$, $1 \le k \le K$, in 2D. Figure 1.10 shows basis functions of four different types of Lagrange multiplier spaces. All four types satisfy the assumptions (Sa)–(Sd). The two pictures on the right represent elements of a dual basis which also satisfy (Se). In each case, we take $W_{0;h_m}(\gamma_m)$ as $\widetilde{W}_{0;h_m}(\gamma_m)$ and choose $\widetilde{W}_{h_m}(\gamma_m)$ equal to the standard Lagrange multiplier space defined by (1.2). The basis functions under consideration are the standard nodal ones. Then, the approximation property (Sc) is satisfied without any further assumptions.

We consider four different Lagrange multiplier spaces for the piecewise linear case denoted by M_h^i, $1 \le i \le 4$, each defining a constrained finite element space V_h^i. The corresponding mortar finite element solutions of (1.14) and (1.15) are denoted by u_h^i and (u_h^i, λ_h^i), respectively. The nodal basis functions of $M_h^i(\gamma_m)$ are denoted by ψ_l^i, $1 \le l \le N_m$, where N_m is the number of vertices x_l in the interior of γ_m. For simplicity, we suppress the index m in the case of the basis functions and the vertices. The enumeration of the vertices x_l is lexicographically, and the two endpoints of γ_m are denoted by x_0 and x_{N_m+1}. Furthermore, the length h_l, $1 \le l \le N_m + 1$, is defined by $h_l := \|x_{l-1} - x_l\|$, and we define the diagonal matrix D by $d_{ii} := h_i$, $1 \le i \le N_m$.

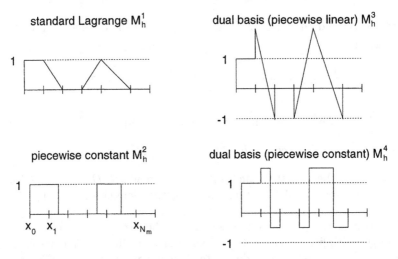

Fig. 1.10. Different types of basis functions for Lagrange multiplier spaces

We set M_h^1 equal to the standard Lagrange multiplier space; see [BMP93, BMP94], and observe that the nodal basis functions are continuous and piecewise linear. The nodal basis functions in the interior of the interface are the standard hat functions, and they are modified only in the neighborhood of the two endpoints; see the upper left of Fig. 1.10. It is then easy to see that the standard Lagrange multiplier space $M_{h_m}^1(\gamma_m)$ and its set of nodal basis functions satisfy the assumptions (Sa) and (Sb). Observing that M_i is symmetric in this special case and that M_i, D_i, D_ψ and $\widetilde{D}_\psi$, D_ψ^3 and $\widetilde{D}_i$, D_i^{-1} are spectrally equivalent, respectively, we find (Sd).

The second space M_h^2 is based on piecewise constant functions. We define the nodal basis functions ψ_i^2 of $M_h^2(\gamma_m)$ by

$$\psi_i^2(x) := \begin{cases} 1, & x \in [\frac{1}{2}(x_{i-1} + x_i), \frac{1}{2}(x_i + x_{i+1})], \\ 0, & \text{elsewhere,} \end{cases} \quad 2 \le i \le N_m - 1 \ .$$

$\psi_1^2(x)$ and $\psi_{N_m}^2(x)$ are equal one on $[x_0, \frac{1}{2}(x_1 + x_2)]$ and $[\frac{1}{2}(x_{N_m-1} + x_{N_m}), x_{N_m+1}]$, respectively, and zero elsewhere; see the lower left of Fig. 1.10. Obviously, the locality of the supports (Sa) and the approximation property (Sb) are satisfied. To verify (Sd), we consider the mass matrices. We find that $(d_\psi)_{ii} = 0.5(h_i + h_{i+1})$, $2 \le i \le N_m - 1$, $(d_\psi)_{11} = 0.5(2h_1 + h_2)$, $(d_\psi)_{N_m N_m} = 0.5(h_{N_m} + 2h_{N_m+1})$, $(d_1)_{ii} = 1/3(h_i + h_{i+1})$, $1 \le i \le N_m$. The mass matrix M_1 in (Sd) is symmetric positive definite and tridiagonal

$$
M_1 = \begin{pmatrix} \frac{1}{2}h_1 + \frac{3}{8}h_2 & \frac{1}{8}h_2 & & & \\ \frac{1}{8}h_2 & \frac{3}{8}(h_2 + h_3) & \frac{1}{8}h_3 & & \\ & \ddots & \ddots & \ddots & \\ & & \frac{1}{8}h_{N_m-1} & \frac{3}{8}(h_{N_m-1} + h_{N_m}) & \frac{1}{8}h_{N_m} \\ & & & \frac{1}{8}h_{N_m} & \frac{3}{8}h_{N_m} + \frac{1}{2}h_{N_m+1} \end{pmatrix}.
$$

The explicit representation of the matrices shows the spectral equivalences of M_i, D_i, D_ψ, $\tilde{D}_i^{-1}$ and D as well as between $\tilde{D}_\psi$ and D^3. These equivalences guarantees (Sd).

The next two examples satisfy the biorthogonality relation (Se), in addition to (Sa)–(Sd). Two different sets of biorthogonal functions are introduced. In the first case, the functions are piecewise linear whereas in the second case they are piecewise constant. The space M_h^3 was originally introduced in [Woh00a] for the mortar setting and is spanned by piecewise linear but discontinuous functions. The main advantage of these basis functions is their biorthogonality with respect to the standard hat functions. Here, we review the definition of the nodal basis functions and define

$$
\psi_i^3(x) := \begin{cases} \frac{1}{h_i}(2\|x - x_{i-1}\| - \|x - x_i\|), & x \in [x_{i-1}, x_i] , \\ \frac{1}{h_{i+1}}(2\|x - x_{i+1}\| - \|x - x_i\|), & x \in [x_i, x_{i+1}] , \\ 0, & \text{elsewhere} , \end{cases}
$$

for $2 \le i \le N_m - 1$. The two basis functions $\psi_1^3(x)$ and $\psi_{N_m}^3(x)$ close to the endpoints of γ_m are equal one on $[x_0, x_1]$ and $[x_{N_m}, x_{N_m+1}]$, respectively, while they have the same structure as the other basis functions elsewhere; see the upper right of Fig. 1.10.

Our next example is also a dual basis, but in contrast to $M_h^3(\gamma_m)$, its basis functions are piecewise constant. For $2 \le i \le N_m - 1$, let

$$
\psi_i^4(x) := \begin{cases} \frac{3}{2}, & x \in [\frac{1}{2}(x_{i-1} + x_i), \frac{1}{2}(x_i + x_{i+1})] , \\ -\frac{1}{2}, & x \in [x_{i-1}, \frac{1}{2}(x_{i-1} + x_i)) \cup (\frac{1}{2}(x_i + x_{i+1}), x_{i+1}] , \\ 0, & \text{elsewhere} , \end{cases}
$$

and let $\psi_1^4(x)$ and $\psi_{N_m}^4(x)$ be equal one on $[x_0, x_1]$ and $[x_{N_m}, x_{N_m+1}]$, respectively, while they have the same structure as the other basis functions elsewhere; see the lower right of Fig. 1.10.

The conditions (Sa), (Sd), and (Se) can be easily verified for the two Lagrange multiplier spaces $M_{h_m}^l(\gamma_m)$, $l \in \{3, 4\}$, associated with the two sets of dual basis functions. In both cases, we find that

$$1 = \sum_{i=1}^{N_m} \psi_i^l, \quad l \in \{3, 4\} \ ,$$

and thus $P_0(\gamma_m) \subset M_{h_m}^l(\gamma_m)$, $l \in \{3, 4\}$. By means of Lemma 1.12, we obtain (Sb). Moreover, we find

$$\int_{\gamma_m} \psi_i^l \theta_j \, d\sigma = \delta_{ij} \int_{\gamma_m} \phi_j \, d\sigma = \delta_{ij} \frac{3}{2} \int_{\gamma_m} \phi_j^2 \, d\sigma, \quad 1 \leq i, j \leq N_m, \ l \in \{3, 4\} \ .$$

To get a better understanding of the Lagrange multiplier spaces based on dual basis functions, we consider, in the rest of this subsubsection, the mortar projection in more detail. Here, we restrict ourselves to the special case of a uniform triangulation of γ_m with $h_m := |e|$ and use $M_{h_m}^3(\gamma_m) = \mathrm{span} \{\psi_i^3 \mid 1 \leq i \leq N_m\}$. Then,

$$\|\Pi_m v\|_{0;e}^2 = \frac{1}{6} h_m \left((\Pi_m v(p_0))^2 + (\Pi_m v(p_1))^2 + (\Pi_m v(p_0) + \Pi_m v(p_1))^2 \right) \ ,$$

where p_0 and p_1 are the two endpoints of e. Using the definition of the mortar projection (1.16) and summing over all elements, we get

$$\|\Pi_m v\|_{0;\gamma_m}^2 \leq \frac{17}{6} \|v\|_{0;\gamma_m}^2 \ .$$

Figure 1.11 shows the trace of a nodal basis function of V_h^3. The trace on the mortar and non-mortar side is illustrated for the two different choices of the mortar side. In the two pictures on the left, the mortar side is associated with the finer mesh whereas in the two pictures on the right, the mortar side is associated with the coarser mesh.

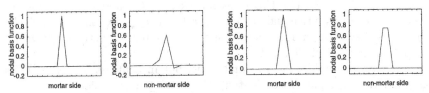

Fig. 1.11. Nodal basis function on a mortar and non-mortar side, (dual)

In contrast to Fig. 1.6, where the same situation for a basis function in V_h^1 is illustrated, the support of the trace is bounded on both sides by a constant times the meshsize. This is the main advantage of using dual basis functions; the locality of the supports of the nodal basis functions of the constrained space is preserved.

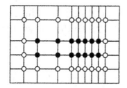 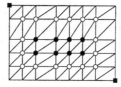

Fig. 1.12. Trace of a hexahedral (left) and a simplicial (right) triangulation

1.2.4.2 The First Order Case in 3D. As a second case, we consider first order Lagrangian finite elements in 3D. Two different situations are discussed separately; see Fig. 1.12. In the first, the triangulation on the interface γ_m is a tensor grid, in the second, it is a simplicial triangulation. For simplicity, we assume that the interfaces are rectangles.

As in the 2D case, the definition of the dual basis functions has to be modified in the neighborhood of the boundary $\partial\gamma_m$. We recall that $W_{h_m}(\gamma_m)$ is the trace space of bilinear or linear finite elements on γ_m, and that ϕ_l is the standard nodal basis function associated with the vertex x_l. In a first step, we define a basis which is biorthogonal to the standard nodal basis of $W_{h_m}(\gamma_m)$. In a second step, we reduce the number of dual basis functions by the number of vertices on the boundary of γ_m without loosing the biorthogonality and the approximation property of the dual space. We define $\widetilde{\psi}_l$ elementwise for $T \in \mathcal{S}_{m;h_m}$, such that supp $\widetilde{\psi}_l =$ supp ϕ_l, by

$$\widetilde{\psi}_{l|_T} := (4\phi_l - 2\phi_{a1(l)} - 2\phi_{a2(l)} + \phi_{o(l)})|_T , \quad T \in \text{supp } \phi_l$$

in the case of a hexahedral triangulation and by

$$\widetilde{\psi}_{l|_T} := (3\phi_l - \phi_{a1(l)} - \phi_{a2(l)})|_T , \quad T \in \text{supp } \phi_l$$

in the case of a simplicial triangulation. The indexing of the nodes, $a1(l)$, $a2(l)$, and $o(l)$, is illustrated in Fig. 1.13. The indices of the two adjacent vertices of x_l on ∂T are denoted by $a1(l)$ and $a2(l)$ and the index of the opposite one, in the case of a hexahedral triangulation, by $o(l)$.

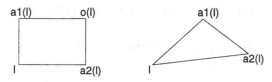

Fig. 1.13. Indices in the case of a rectangle and a triangle

The basis functions ψ_l are associated with the interior vertices x_l of γ_m. We now set $\psi_l := \widetilde{\psi}_l$ in the interior of γ_m, i.e., if $\partial(\text{supp}\phi_l) \cap \partial\gamma_m = \emptyset$. The corresponding vertices are marked by filled circles in Fig. 1.12. For all other vertices $x_l \in \gamma_m$, we have to modify the definition. The corresponding vertices are marked by empty circles in Fig. 1.12. To do so, we introduce two sets of vertices $\mathcal{M}$ and $\mathcal{M}_l$ given by

$$\mathcal{M}_l := \{x_j \mid x_j \in \partial\gamma_m \cap \partial(\mathrm{supp}\phi_l)\}, \quad \mathcal{M} := \{x_j \mid x_j \in \partial\gamma_m\} \setminus \bigcup_{x_l \in \gamma_m} \mathcal{M}_l .$$

The set $\mathcal{M}$ can be non-empty only in the case of simplicial triangulations; it is always empty in the case of a tensor grid. In the situation, which is shown in the right part of Fig. 1.12, $\mathcal{M}$ contains two elements associated with the upper left and the lower right corner, marked with filled squares. The modified dual basis functions are given by

$$\psi_l := \widetilde{\psi}_l + \sum_{x_j \in \mathcal{M}_l} \frac{1}{n_j} \widetilde{\psi}_j + \sum_{x_j \in \mathcal{M}} \alpha_{lj} \widetilde{\psi}_j, \quad x_l \in \gamma_m . \tag{1.31}$$

Here, n_j is the number of sets $\mathcal{M}_k$ such that $x_j \in \mathcal{M}_k$, and $\alpha_{lj} = 1$ if $\partial(\mathrm{supp}\phi_l) \cap \partial(\mathrm{supp}\,\phi_j)$ contains an edge and zero elsewhere. We refer to [BM97] for an introduction of a standard Lagrange multiplier space, and to [BD98] for some more sophisticated choices of the weights in (1.31). In particular, it is possible to replace $1/n_j$ by a weight involving the areas of the adjacent elements.

Obviously (Sa) is satisfied. To verify (Sc) and (Sd), we have to specify the spaces $\widetilde{W}_{0;h_m}(\gamma_m)$ and $\widetilde{W}_{h_m}(\gamma_m)$ and their basis functions. As in the 2D case, we set $\widetilde{W}_{0;h_m}(\gamma_m) := W_{0;h_m}(\gamma_m)$ and choose the standard hat functions as basis functions $\theta_i := \phi_i$. The space $\widetilde{W}_{h_m}(\gamma_m)$ is obtained from $\widetilde{W}_{0;h_m}(\gamma_m)$ by modifying the standard hat functions in exactly the same way as the Lagrange multiplier basis functions before:

$$\widetilde{\theta}_l := \phi_l + \sum_{x_j \in \mathcal{M}_l} \frac{1}{n_j} \phi_j + \sum_{x_j \in \mathcal{M}} \alpha_{lj} \phi_j, \quad x_l \in \gamma_m .$$

It is now easy to see that $\sum_{x_l \in \gamma_m} \psi_l = \sum_{x_l \in \gamma_m} \widetilde{\theta}_l = 1$. Thus (Sc) is satisfied and $P_0(\gamma_m) \subset M_{h_m}(\gamma_m)$. Furthermore, a straightforward computation shows that the following biorthogonality relation holds

$$\int_{\gamma_m} \psi_l \, \theta_k \, d\sigma = \int_{\gamma_m} \psi_l \, \phi_k \, d\sigma = \delta_{lk} \int_{\gamma_m} \phi_k \, d\sigma, \quad x_l, x_k \in \gamma_m .$$

This implies (Se) and (Sd) for $i = 1$. (Sd), for $i = 2$, is obtained by observing that M_2, D_1 and D_2 are spectrally equivalent as well as $\widetilde{D}_1$ and $\widetilde{D}_2$. Applying Lemma 1.12 yields (Sb). As in the 2D case, the biorthogonality relation results in a mass matrix, on the non-mortar side, which has a diagonal block associated with the interior nodes. In addition, the nodal basis functions of

the constrained space have local support. We remark that we do not have
supp ψ_l = supp ϕ_l in the neighborhood of the boundary of γ_m.

In the special situation of a tensor product mesh on a rectangular interface, the construction of a dual basis function can be simplified. Piecewise bilinear and piecewise constant dual basis functions can be defined as
a product in terms of the dual basis functions ψ_i^3 and ψ_i^4 given in the previous subsubsection, respectively. Then, the support of this simplified dual
basis function ψ_l^s associated with the vertex $x_l \in \gamma_m$ is the union of the four
adjacent rectangles sharing the vertex x_l. Figure 1.14 illustrates the isolines
of such a dual basis function if $\partial(\mathrm{supp}\,\psi_l^s) \cap \partial\gamma_m = \emptyset$.

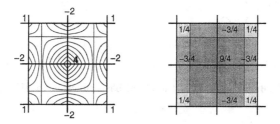

Fig. 1.14. Isolines of piecewise bilinear and piecewise constant dual basis functions

Comparing these simplified piecewise bilinear dual basis functions with
the definition (1.31), we find that they are the same if $\partial(\mathrm{supp}\,\psi_l^s) \cap \partial\gamma_m = \emptyset$.
However, in the neighborhood of the boundary of γ_m, we observe a differ-
ence. In particular, the tensor product dual basis functions ψ_l^s have a smaller
support than ψ_l. We distinguish between three different types of vertices x_l.
The inner ones, i.e., $\partial(\mathrm{supp}\,\phi_l) \cap \partial\gamma_m = \emptyset$, are marked with empty squares,
the ones close to the corners, i.e., $\partial(\mathrm{supp}\,\phi_l)$ contains one corner of γ_m, by
empty circles and all other vertices are marked by filled circles; see Fig. 1.15.

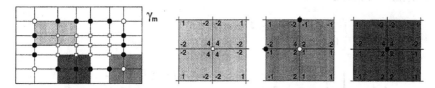

Fig. 1.15. Different types of piecewise bilinear dual basis functions

Figure 1.15 shows the different groups of vertices and one piecewise bilinear dual basis function ψ_l^s for each vertex type. The other ones can be
obtained by local rotations. For the simplicial case, it is also possible to
modify the definition (1.31) in the neighborhood of the boundary of γ_m; see
[KLPV00]. Then, the nodal basis function ϕ_l and the dual basis ψ_l^s have the
same support. The advantage of the simplified dual basis functions is the

smaller support. In contrast to supp ψ_l, the support of ψ_l^s contains for all indices l only the elements sharing the node x_l. The assumptions (Sa)–(Se) are easy to verify.

1.2.4.3 The Second Order Case in 2D.
In our last example, we consider the case $n_k = 2$ in 2D. Again, we set $\widetilde{W}_{0;h_m}(\gamma_m) := W_{0;h_m}(\gamma_m)$, and $\widetilde{W}_{h_m}(\gamma_m)$ is defined as the standard Lagrange multiplier space (1.2) with $n_k = 2$. Both spaces are associated with the set of the corresponding nodal basis functions. Figure 1.16 illustrates the numbering of the nodal basis functions θ_i, $1 \leq i \leq N_m$, in $\widetilde{W}_{0;h_m}(\gamma_m)$.

$$\theta_1 \quad \theta_2 \qquad\qquad\qquad\qquad \theta_{N_m-1} \ \theta_{N_m}$$

Fig. 1.16. Numbering of the nodal basis functions of $\widetilde{W}_{0;h_m}(\gamma_m)$

The dual basis functions are now given in terms of θ_i, $1 \leq i \leq N_m$. We define ψ_i, for $i = 2l + 1$, $1 \leq l \leq (N_m - 3)/2$, by

$$\psi_i(x) := \begin{cases} (-1 + \frac{5}{2}\theta_i)(x), & x \in \text{supp}\,\theta_i \ , \\ 0, & \text{elsewhere} \ , \end{cases}$$

and for $i = 2l$, $2 \leq l \leq (N_m - 3)/2$ by

$$\psi_i(x) := \begin{cases} (\frac{1}{2} - \frac{3}{4}\theta_{i-1} + \theta_i - \frac{3}{4}\theta_{i+1})(x), & x \in \text{supp}\,\theta_i \ , \\ 0, & \text{elsewhere} \ . \end{cases}$$

The two first basis functions, ψ_1, ψ_2, and the two last ones , ψ_{N_m-1}, ψ_{N_m}, are defined differently by

$$\psi_1(x) := \begin{cases} \frac{2}{h_1}\|x - x_1\|, & x \in \text{supp}\,\theta_1 \ , \\ 0, & \text{elsewhere} \ , \end{cases}$$

$$\psi_2(x) := \begin{cases} \frac{1}{h_1}(\|x - x_0\| - \|x - x_1\|), & x \in \text{supp}\,\theta_1 \ , \\ (\frac{1}{2} + \theta_2 - \frac{3}{4}\theta_3)(x), & x \in \text{supp}\,\theta_2 \setminus \text{supp}\,\theta_1 \ , \\ 0, & \text{elsewhere} \ , \end{cases}$$

with ψ_{N_m-1} and ψ_{N_m} given in a similar way.

A straightforward computation shows that (Se) is satisfied. In this case, the spectral equivalence (Sd) follows from the spectral equivalence of D_ψ and D_1. Furthermore, we find $\sum_{i=1}^{N_m} \psi_i = 1$. Observing that $\psi_{2i} + 0.5\psi_{2i-1} + 0.5\psi_{2i+1}$, is the standard piecewise linear and continuous hat function associated with the vertex x_i, $1 \leq i \leq (N_m - 1)/2$, the approximation property

(Sb) follows from Lemma 1.12. Figure 1.17 illustrates the dual basis functions ψ_i, $1 \leq i \leq N_m$.

We observe that a priori estimates for the Lagrange multiplier of order h^n can be obtained by using piecewise polynomials of order $(n-1)$ for the Lagrange multiplier; see also [SS98]. Since the a priori estimates for the Lagrange multiplier depends on the estimate for $u - u_h$, the a priori estimate for the flux cannot be improved by choosing higher order elements for the Lagrange multiplier.

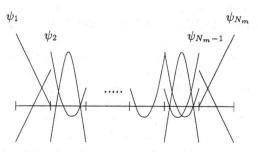

Fig. 1.17. Dual basis functions, ($n_k = 2$)

Remark 1.13. *The assumptions (Sa)–(Sd) also allow us to work with Lagrange multipliers which are defined on a coarser triangulation than $\mathcal{S}_{m;h_m}$. For example if $\mathcal{S}_{m;h_m}$ is obtained by a uniform refinement step from $\mathcal{S}_{m;2h_m}$, then $M_{2h_m}(\gamma_m)$ satisfies (Sa)–(Sd) if $M_{h_m}(\gamma_m)$ does.*

Finally, we note that the concept of dual Lagrange multiplier spaces can be generalized to higher order elements. The construction of higher order dual basis functions is very technical, and we refer the interested reader to [OW00]. In [OW00], dual basis functions are constructed for conforming P_n-elements in 2D. The number of edges contained in the support of a dual basis function is at most three and independent of the order n. Moreover, the assumptions (Sa)–(Se) are satisfied by construction. Thus, optimal higher order mortar methods for dual Lagrange multiplier spaces are obtained.

1.3 Discretization Techniques Based on the Product Space

In the previous subsections, two possible approaches to the implementation of a mortar discretization have been discussed. The first gives rise to a positive definite problem, and is based on the constrained space. The second works with the unconstrained product space and a Lagrange multiplier space and results in a saddle point formulation, where the weak continuity is enforced by means of the Lagrange multiplier space. In this subsection, we consider discretization techniques working only on the unconstrained product space. Such a possibility is also discussed in [BH99, Ste98]. An idea introduced by Nitsche in [Nit70] is rediscovered and generalized to non-matching triangulations. It turns out to be a penalty method where the penalty parameter does not depend on the meshsize and does not influence the condition number of the resulting linear system. We refer to [BH99] for details and some numerical results including a posteriori error estimators.

Here, we develop a different idea. We note that the numerical solution of the mortar method is, very often, based on its saddle point formulation [AK95, AKP95, AMW96, AMW99, BD98, BDH99b, BDW99, EHI$^+$98, EHI$^+$00, HIK$^+$98, Kuz95a, Kuz95b, Kuz98, KW95, Lac98, LSV94, WW98, WW99]. Working with this indefinite formulation has the advantage that the unconstrained product spaces associated with a nested sequence of global triangulations are nested, while the nonconforming constrained spaces are not. Our approach is based on this observation, and we introduce an unsymmetric Dirichlet–Neumann coupling and a symmetric and positive definite variational problem on the unconstrained product space. We then show that the solution of these problems coincides with the mortar finite element solution. The idea is not restricted to first order discretizations or to 2D.

We assume that the pair (X_h, M_h) defines a constrained space V_h such that the nonconforming variational problem (1.14) and the saddle point problem (1.15) are well defined and yield optimal a priori bounds for the discretization errors, see conditions (Sa)–(Sd) in Sect. 1.2. Then, we have only to ensure that the nodal basis functions of $M_{h_m}(\gamma_m)$ and $\widetilde{W}_{0;h_m}(\gamma_m)$ form a biorthogonal set

$$\int_{\gamma_m} \psi_l \theta_k \, d\sigma = c_l \delta_{lk} \|\theta_k\|^2_{0;\gamma_m}, \quad 1 \le l, k \le N_m \, , \qquad (1.32)$$

where $c \le c_l \le C$, $1 \le l \le N_m$, and $N_m = \dim M_{h_m}(\gamma_m) = \dim \widetilde{W}_{0;h_m}(\gamma_m)$, see condition (Se) in Sect. 1.2. We recall that $\widetilde{W}_{0;h_m}(\gamma_m)$ is a suitable subspace of the finite element trace space on the non-mortar side with zero value on the boundary of γ_m. The index m is dropped for the basis functions and also for the constants c_l. Examples of such pairings have been given in the previous subsections for the first order cases in 2D and in 3D, and for the quadratic case in 2D. We recall that the only examples introduced in the previous subsections which do not satisfy these conditions are the standard Lagrange multiplier space M_h^1 and the space M_h^2. Here, we restrict ourselves to the special case $N_m = \dim W_{0;h_m}(\gamma_m)$ and $\theta_l = \phi_l$, $1 \le l \le N_m$, where ϕ_l are the standard nodal basis functions.

Our interest in the introduction of a new, third, mortar formulation is based on the following observation: Saddle point problems, like (1.15), are, very often, solved by Uzawa like algorithms, and inner and outer iterations are required. Recently such a multigrid technique has been studied in [WW99] for the traditional mortar approach. In this case, a modified Schur complement system has to be solved iteratively in each smoothing step. This method is a generalization of the techniques given in [BD98, BDW99, Bra01, WW98] where the modified Schur complement system was solved exactly. Another approach is analyzed in [Kuz95a, Kuz95b] and further used in [EHI$^+$98, EHI$^+$00, HIK$^+$98, Kuz98]. A good preconditioner for the exact Schur complement is required, and the iterative solution of the saddle point problem is obtained by a generalized Lanczos method.

It is also true that standard multigrid techniques cannot be applied for the positive definite formulation (1.14) which is associated with a nonconforming space. We refer to [AMW96, AMW99, BDH99b, CDS98, CW96, Dry96, Dry97, Dry98a, Dry98b, GP00] for multilevel and multigrid methods and domain decomposition techniques. Recently, multigrid methods for the nonconforming formulation have been developed [BDH99b, GP00] which involve the solution of a mass matrix system in each smoothing step. Dirichlet–Neumann type preconditioners have been proposed for the positive definite system in [Dry99, Dry00]; see also [LSV94].

We now introduce a third equivalent mortar formulation. It is based on the unconstrained product space and gives rise to a symmetric positive definite formulation on which we can apply standard multigrid techniques. The rest of this section concerns the construction of a symmetric positive definite system defined on the product space. We present a variational formulation on the product space as well as its algebraic form. The stiffness matrix associated with the unconstrained product space can easily be obtained by eliminating the Lagrange multiplier. The starting point is the observation that the Lagrange multiplier λ_h is given in a postprocessing step. But it can be easily evaluated in terms of the right hand side and the mortar solution u_h only in the case that the biorthogonality relation (1.32) holds. Otherwise, the elimination involves the inverse of a mass matrix. We introduce an abstract framework for a new equivalent mortar setting based on the unconstrained product space. The analysis of our system on the product space is carried out using the tools of the mortar framework.

The new idea is to use a biorthogonal basis for the Lagrange multiplier space and to eliminate the Lagrange multiplier in the saddle point formulation. We begin by considering, in Subsect. 1.3.1, a Dirichlet–Neumann coupling for the special case of two subdomains. Subsections 1.3.2 and 1.3.3 concern the construction of a symmetric positive definite system defined on the product space. Its condition number is bounded by a constant times $1/h^2$, as in the usual finite element context. We present the variational formulation on the product space in Subsect. 1.3.2 and its algebraic form in Subsect. 1.3.3.

1.3.1 A Dirichlet–Neumann Formulation

Let us consider the following special situation: The domain $\Omega \subset \mathbb{R}^2$ is decomposed into two nonoverlapping subdomains, $\overline{\Omega} = \overline{\Omega}_1 \cup \overline{\Omega}_2$, and meas $(\partial\Omega \cap \partial\Omega_i) > 0$, $i \in \{1, 2\}$; see Fig. 1.18.

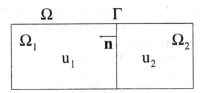

Fig. 1.18. Decomposition into two subdomains

On Ω_1, we consider a Dirichlet boundary value problem

$$
\begin{aligned}
-\mathrm{div}(a\nabla u_1) + bu_1 &= f, && \text{in } \Omega_1 \ , \\
u_1 &= 0, && \text{on } \Gamma_1 := \partial\Omega \cap \partial\Omega_1 \ , \\
u_1 &= g_D, && \text{on } \Gamma := \mathrm{int}(\partial\Omega_1 \cap \partial\Omega_2) \ ,
\end{aligned}
\tag{1.33}
$$

where $g_D \in H_{00}^{1/2}(\Gamma)$. Its variational formulation is: Find $u_1 \in H_{g_D;\Gamma_1}^1(\Omega_1)$ such that

$$
a_1(u_1, v_1) = (f, v_1)_{0;\Omega_1}, \quad v_1 \in H_0^1(\Omega_1) \ .
$$

Here, $H_{g_D;\Gamma_1}^1(\Omega_1)$ is the subset of $H^1(\Omega_1)$ with a vanishing trace on Γ_1 and a trace equal to g_D on Γ. The bilinear forms $a_i(\cdot, \cdot)$, $1 \le i \le 2$, are given by

$$
a_i(v, w) := \int_{\Omega_i} a\nabla v \nabla w + bvw \, dx, \quad v, w \in H^1(\Omega_i) \ .
$$

On Ω_2, we solve a partial differential equation with a Neumann boundary condition on Γ

$$
\begin{aligned}
-\mathrm{div}(a\nabla u_2) + bu_2 &= f, && \text{in } \Omega_2 \ , \\
u_2 &= 0, && \text{on } \Gamma_2 := \partial\Omega \cap \partial\Omega_2 \ , \\
a\frac{\partial u_2}{\partial \mathbf{n}} &= g_N, && \text{on } \Gamma \ ,
\end{aligned}
\tag{1.34}
$$

where $\mathbf{n}$ denotes the outer unit normal on Ω_2. The weak variational formulation is given by: Find $u_2 \in H_{\Gamma_2}^1(\Omega_2)$ such that

$$
a_2(u_2, v_2) = (f, v_2)_{0;\Omega_2} + \int_{\Gamma} g_N v_2 \, d\sigma =: f_2(v_2), \quad v_2 \in H_{\Gamma_2}^1(\Omega_2) \ .
$$

Here, $H_{\Gamma_i}^1(\Omega_i)$, $1 \le i \le 2$, denotes the subspace of $H^1(\Omega_i)$ having a vanishing trace on the outer boundary Γ_i.

Let us assume for the moment that the solution u of (1.1) and the flux on the interface Γ are known. Then, the choice $g_D := u|_\Gamma$ and $g_N := a\partial u/\partial \mathbf{n}|_\Gamma$ yields $u_1 = u|_{\Omega_1}$ and $u_2 = u|_{\Omega_2}$.

The discrete spaces of conforming P_1-Lagrangian finite elements on Ω_1 and Ω_2, associated with the simplicial triangulations $\mathcal{T}_1$ and $\mathcal{T}_2$, are denoted by $X_1 \subset H^1_{\Gamma_1}(\Omega_1)$ and $X_2 \subset H^1_{\Gamma_2}(\Omega_2)$, respectively. The set of corresponding vertices are denoted by $\mathcal{P}_1$ and $\mathcal{P}_2$, and the subsets of vertices on the interface by $\mathcal{P}^\Gamma_1$ and $\mathcal{P}^\Gamma_2$. Then, (1.33) and (1.34) can be discretized by means of the discrete spaces X_1 and X_2, respectively. We will now use the same notation for an element in X_i, $1 \leq i \leq 2$, and its vector representation with respect to the nodal basis, $\{\phi_p,\ p \in \mathcal{P}_i\}$, of X_i.

The matrix representation of the Dirichlet problem (1.33) can be given by

$$A^1_D u_{h;1} := \begin{pmatrix} A^1_{II} & A^1_{I\Gamma} \\ 0 & \mathrm{Id} \end{pmatrix} \begin{pmatrix} u^1_I \\ u^1_\Gamma \end{pmatrix} = \begin{pmatrix} f^1_I \\ f^1_\Gamma \end{pmatrix} . \tag{1.35}$$

Here, A^1_{II} and $A^1_{I\Gamma}$ are the stiffness matrices associated with the bilinear form $a_1(\cdot, \cdot)$ restricted to $X^1_I \times X^1_I$ and $X^1_I \times X^1_\Gamma$, respectively. The space X^1_Γ is spanned by the nodal basis functions ϕ_p, $p \in \mathcal{P}^\Gamma_1$, and $X^1_I := X_1 \cap H^1_0(\Omega_1)$. The right hand side f^1_I is associated with the linear form $\int_{\Omega_1} fv\, dx$, and f^1_Γ depends on g_D.

Dirichlet boundary conditions are often realized by a pointwise equality at the nodes of the triangulation

$$(f^1_\Gamma)_p = g_D(p), \quad p \in \mathcal{P}^\Gamma_1 ,$$

if the boundary data function g_D is continuous. Here, we use a different approach and specify the Dirichlet boundary condition by a weak integral condition

$$(f^1_\Gamma)_p := \frac{2}{|\mathrm{supp}\ \phi_p|} \int\limits_{\mathrm{supp}\ \phi_p} \psi_p g_D\, d\sigma, \quad p \in \mathcal{P}^\Gamma_1 ,$$

where $\psi_p := 3\phi_p - 1$. We refer to [SZ90], for a discussion of nonhomogeneous Dirichlet boundary conditions, and observe that ψ_p is the same as a nodal basis function of the dual space $M^3_{h_m}(\gamma_m)$ in Sect. 1.2, if the vertex p is not next to an endpoint of the interface.

In contrast to Dirichlet boundary conditions, Neumann boundary conditions g_N are natural and enter directly into the variational formulation. The discrete variational formulation of (1.34) can be written as

$$A^2_N u_{h;2} := \begin{pmatrix} A^2_{II} & A^2_{I\Gamma} \\ A^2_{\Gamma I} & A^2_{\Gamma\Gamma} \end{pmatrix} \begin{pmatrix} u^2_I \\ u^2_\Gamma \end{pmatrix} = \begin{pmatrix} f^2_I \\ f^2_\Gamma \end{pmatrix} . \tag{1.36}$$

We use the same notation as before; and define $\hat{f}^2_\Gamma$ by

$$(\hat{f}_\Gamma^2)_p := (f_\Gamma^2)_p + \int_\Gamma g_N \phi_p \, d\sigma, \quad (\hat{f}_\Gamma^2)_p := (f, \phi_p)_{0;\Omega_2}, \quad p \in \mathcal{P}_2^\Gamma \ .$$

Unfortunately, the solution and the flux restricted to Γ are, in general, unknown, and the boundary data g_D and g_N are not available. However, if in the continuous setting we set $g_D := u_2$ and $g_N := a\partial u_1/\partial \mathbf{n}$, (1.33) and (1.34) form a coupled system of boundary value problems. Then, $f_2(v_2)$ can be rewritten, and we find that

$$f_2(v_2) = (f, v_2)_{0;\Omega_2} + \int_\Gamma a\frac{\partial u_1}{\partial \mathbf{n}} v_2 \, d\sigma = -a_1(u_1, v) + (f, v)_0 \ .$$

Here, $v \in H_0^1(\Omega)$ is an extension of $v_2 \in H_{\Gamma_2}(\Omega_2)$. We remark that the jump of the trace of v across Γ is zero. However, the jump of the flux of v does not have to be zero.

These observations form the starting point for our discrete approach. When working with non-matching triangulations at the interface Γ, it is in general not possible to extend a function $v_{h;2} \in X_2$ to $X_1 \times X_2$ such that the jump of the trace is zero. However, in the mortar approach, it is standard to replace the strong continuity of the jump by a weaker one. Observing that each element in $X_1 \times X_2$ is uniquely defined by its values at the vertices $p_1 \in \mathcal{P}_1$ and $p_2 \in \mathcal{P}_2$, a suitable discrete extension $Ev_{h;2}$ of $v_{h;2} \in X_2$ is constructed in the following way:

$$Ev_{h;2}(p) := \begin{cases} v_{h;2}(p), & p \in \mathcal{P}_2 \ , \\ 0, & p \in \mathcal{P}_1 \setminus \mathcal{P}_1^\Gamma \ , \\ \frac{2}{|\text{supp } \phi_p|} \int_{\text{supp}\phi_p} \psi_p v_{h;2} \, d\sigma, & p \in \mathcal{P}_1^\Gamma \ . \end{cases} \tag{1.37}$$

We note that the extension operator E is not uniformly bounded in the broken H^1-norm.

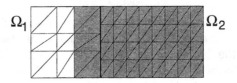

Ω_1 Ω_2

Fig. 1.19. Support of the extended function

The shadowed region in Fig. 1.19 illustrates the structure of the support of $Ev_{h;2} \in X_1 \times X_2$. Using the definition of ψ_p and E, we find

$$\int_{\text{supp}\phi_p} \psi_p v_{h;2} \, d\sigma = \int_{\text{supp}\phi_p} \psi_p (Ev_{h;2})_{|\Omega_2} \, d\sigma = \int_{\text{supp}\phi_p} \psi_p (Ev_{h;2})_{|\Omega_1} \, d\sigma \ .$$

We are now ready to formulate our discrete approach: Find $(u_{h;1}, u_{h;2}) \in X_1 \times X_2$ such that

$$
\begin{array}{ll}
a_1(u_{h;1}, v_{h;1}) = (f, v_{h;1})_{0;\Omega_1}, & v_{h;1} \in X_1^I, \\
u_{h;1}(p) = E u_{h;2}(p), & p \in \mathcal{P}_1^\Gamma, \\
a_2(u_{h;2}, v_{h;2}) = (f, E v_{h;2})_0 - a_1(u_{h;1}, E v_{h;2}), & v_{h;2} \in X_2,
\end{array}
\tag{1.38}
$$

where $E v_{h;2}$ and $E u_{h;2}$ are the discrete extensions of $v_{h;2}$ and $u_{h;2}$ given by (1.37). The solution of (1.38) depends on the definition of E on the interface but not on the values at p, $p \in \mathcal{P}_1 \setminus \mathcal{P}_1^\Gamma$. Thus the trivial extension E by zero into the interior of Ω_1 could be replaced by a discrete harmonic extension without influencing the solution of (1.38).

The matrix formulation of (1.38) reflects directly the Dirichlet–Neumann coupling between the two subdomains

$$
\begin{pmatrix}
A_{II}^1 & A_{I\Gamma}^1 & 0 & 0 \\
0 & \mathrm{Id} & -M^T & 0 \\
M A_{\Gamma I}^1 & M A_{\Gamma\Gamma}^1 & A_{\Gamma\Gamma}^2 & A_{\Gamma I}^2 \\
0 & 0 & A_{I\Gamma}^2 & A_{II}^2
\end{pmatrix}
\begin{pmatrix}
u_I^1 \\
u_\Gamma^1 \\
u_\Gamma^2 \\
u_I^2
\end{pmatrix}
=
\begin{pmatrix}
f_I^1 \\
0 \\
f_\Gamma^2 + M f_\Gamma^1 \\
f_I^2
\end{pmatrix},
\tag{1.39}
$$

where $M := (m_{p_2 p_1})_{p_2 \in \mathcal{P}_2^\Gamma, p_1 \in \mathcal{P}_1^\Gamma}$ is a scaled mass matrix defined by

$$
m_{p_2 p_1} := \frac{2}{|\operatorname{supp} \phi_{p_1}|} \int_{\operatorname{supp} \phi_{p_1}} \psi_{p_1} \phi_{p_2} \, d\sigma .
$$

Before we analyze the unique solvability of (1.38) and the discretization error in the new setting, we consider our approach in the more general situation with several subdomains and crosspoints. We remark that we have to modify the approach in the neighborhood of a crosspoint. Using the same notations as in Sect. 1.2, we find that the choice of the Lagrange multiplier space M_h plays an essential role. We show in the next subsection that the discrete variational problem on the product space can be obtained from the saddle point problem by locally eliminating the Lagrange multiplier.

Remark 1.14. *We observe that the stiffness matrix in (1.39) reflects the Dirichlet–Neumann coupling between the two subdomains. Applying a block Gauß–Seidel method on this system results in a Dirichlet–Neumann type pre-conditioner. This method is well known for conforming finite element methods; see [SBG96] and the references therein, and has been also studied for the standard mortar approach; see [Dry99, Dry00, LSV94]. For details, we refer to Sect. 2.4.*

1.3.2 Variational Formulations

In this subsection, we introduce our new abstract variational formulation on the unconstrained product space. The only necessary condition is the biorthogonality relation (1.32). An unsymmetric formulation as well as a symmetric one are discussed. A careful analysis shows that the unsymmetric one for the special pairing (X_h, M_h^3) is the Dirichlet–Neumann coupling described in the previous subsection. The symmetric one is positive definite, and as we will see in Sect. 2.3, a standard multigrid method can be applied for the resulting linear system.

In the rest of this section, we assume that the biorthogonality relation (1.32) is satisfied. At first glance the saddle point problem (1.15) has the same structure for all Lagrange multiplier spaces. However, there is an essential difference. Given the solution u_h of (1.14), the Lagrange multiplier $\lambda_{h|_{\gamma_m}}$ can be obtained by solving a mass matrix system. This postprocessing step involves the inverse of a mass matrix which is, in general, dense on each interface γ_m. Only in the case that the biorthogonality relation (1.32) holds will the inverse mass matrix reduce to a diagonal one. Then, the value of an element $v \in V_h$ at a point on one non-mortar side is determined by its values in a small neighborhood of that point on the adjacent mortar side. In particular, the mass matrix associated with the non-mortar side is replaced by a diagonal one reflecting the biorthogonality.

On each non-mortar side γ_m, we can write λ_h as $\lambda_{h|_{\gamma_m}} = \sum_{i=1}^{N_m} \lambda_{m;i}\psi_{m;i}$. Compared with Sect. 1.2, the additional index m indicates the corresponding non-mortar side γ_m. The coefficients $\lambda_{m;i}$ are given in terms of u_h and f by

$$\lambda_{m;i} = \frac{(f, \phi_{m;i})_0 - a(u_h, \phi_{m;i})}{\int_{\gamma_m} \psi_{m;i}\phi_{m;i}\, d\sigma} \quad , \tag{1.40}$$

where $\phi_{m;i}$ is the nodal basis function which has its support in $\Omega_{n(m)}$.

Let us now define linear functionals $g_1 : X_h \longrightarrow M_h$ and $g_2 : X_h \longrightarrow X_h$. Both functionals are associated with the interfaces. We define

$$g_1(v) := \sum_{m=1}^{M} \sum_{i=1}^{N_m} \frac{v|_{\Omega_{n(m)}}(p_{m;i})}{\int_{\gamma_m} \psi_{m;i}\phi_{m;i}\, d\sigma} \psi_{m;i}, \quad v \in X_h \ ,$$

where $p_{m;i}$ are the nodal points on γ_m, i.e., $\phi_{m;i}(p_{m;j}) = \delta_{ij}$. We furthermore use a linear functional $g_2(\cdot)$ which vanishes on V_h. It is defined by

$$g_2(v) := \sum_{m=1}^{M} \sum_{i=1}^{N_m} \frac{b(v, \psi_{m;i})}{\int_{\gamma_m} \psi_{m;i}\phi_{m;i}\, d\sigma} \phi_{m;i}, \quad v \in X_h \ . \tag{1.41}$$

It is easy to see that $g_2(v)$ is only supported in a small neighborhood of the non-mortar side and vanishes on the mortar sides of the interfaces.

Proposition 1.15. *The kernel of $g_2(\cdot)$ is exactly V_h and $g_2(g_2(v)) = g_2(v)$. Moreover, the L^2-norm of $g_2(v)$ on Ω can be bounded by the L^2-norm on the interfaces of the jump of v*

$$\|g_2(v)\|_0^2 \le C \sum_{m=1}^{M} \sum_{e \in S_{m;h_m}} h_e \|[v]\|_{0;e}^2 \ .$$

Proof. The definition of $g_2(\cdot)$ and the biorthogonality relation (1.32) yields the first two assertions. A straightforward calculation shows

$$\|g_2(v)\|_0^2 \le \sum_{m=1}^{M} \sum_{i=1}^{N_m} \left(\frac{b(v, \psi_{m;i})}{\int_{\gamma_m} \psi_{m;i} \phi_{m;i} \, d\sigma} \right)^2 \|\phi_{m;i}\|_0^2$$

$$\le \sum_{m=1}^{M} \sum_{i=1}^{N_m} \|[v]\|_{0;\mathrm{supp}\psi_{m;i}}^2 \, h_{\psi_{m;i}} \ .$$

Here, we have used the assumptions (Sd)–(Se) which guarantee the equivalence of $\|\phi_{m;i}\|_{0;\gamma_m}$ and $\|\psi_{m;i}\|_{0;\gamma_m}$, and of $h_{\psi_{m;i}}$ and $h_{\phi_{m;i}}$. Finally, (Sa) yields that the maximal number of $\psi_{m;i}$ having an overlapping support is bounded independently of the local meshsize.   □

There are now two possibilities to show that an element $v \in X_h$ is also in V_h. The first uses the definition (1.13) of V_h; it is sufficient to prove that $b(v, \psi_{m;i}) = 0$ for $1 \le m \le M$, $1 \le i \le N_m$. The second possibility is based on the projection $g_2(\cdot)$; $v \in X_h$ is an element of V_h if and only if $g_2(v) = 0$.

The following lemma shows an equivalence between the positive definite variational problem (1.14) on the constrained space V_h and a non-symmetric and a symmetric variational problem on the unconstrained product space X_h. The idea behind the introduction of the new variational problem is to use the decomposition $v = (v - g_2(v)) + g_2(v)$. Then, $v - g_2(v)$ is an element of the constrained space V_h.

Lemma 1.16. *Let $u_h \in V_h$ be the unique solution of (1.14). Then, u_h is the unique solution of the non-symmetric variational problem*

$$a(u_h, v - g_2(v)) + b(u_h, g_1(v)) = (f, v - g_2(v))_0, \quad v \in X_h \ , \qquad (1.42)$$

and also of the positive definite symmetric variational problem

$$a(u_h - g_2(u_h), v - g_2(v)) + a(g_2(u_h), g_2(v)) = (f, v - g_2(v))_0, \quad v \in X_h \ . \qquad (1.43)$$

Proof. In a first step, we show that the solution u_h of (1.14) satisfies (1.42) and (1.43). Using that $u_h \in V_h$, we find $b(u_h, g_1(v)) = 0$ and $g_2(u_h) = 0$. Since $g_2(\cdot)$ is a projection, we have $g_2(v - g_2(v)) = 0$ and thus $v - g_2(v) \in V_h$. Combining these observations, we immediately find that u_h satisfies (1.42) and (1.43).

To obtain the unique solvability of (1.42) and (1.43), it is sufficient to show that each solution of (1.42) and (1.43), respectively, is an element of V_h and thus a solution of (1.14). An element $v \in X_h$ is in V_h if and only if $b(v, \mu) = 0$, $\mu \in M_h$, or equivalently if $g_2(v) = 0$. Let us assume that u_h

satisfies (1.42). For each $1 \leq m \leq M$, $1 \leq i \leq N_m$, we have $g_2(\phi_{m;i}) = \phi_{m;i}$ and $g_1(\phi_{m;i}) = \alpha_{m;i}\psi_{m;i}$ with $\alpha_{m;i} \neq 0$. The choice $v := \phi_{m;i}$ yields

$$b(u_h, \psi_{m;i}) = 0 \; ,$$

and therefore u_h is the unique solution of (1.14).

To prove that $u_h \in V_h$ if u_h is the solution of (1.43), we use the second possibility and show that $g_2(u_h) = 0$. Setting $v := g_2(u_h)$ yields

$$a(g_2(u_h), g_2(u_h)) = 0 \; .$$

Since the bilinear form $a(\cdot, \cdot)$ is coercive on $g_2(X_h)$, we find that $g_2(u_h) = 0$.

It is now easy to see that the variational problem (1.43) is symmetric and positive definite. We define $\hat{a}(v, w) := a(v - g_2(v), w - g_2(w)) + a(g_2(v), g_2(w))$, $v, w \in X_h$. Then $\hat{a}(\cdot, \cdot)$ is a symmetric bilinear form satisfying $\hat{a}(v, v) \geq 0$. For each $v \in X_h$ satisfying $\hat{a}(v, v) = 0$, we find that $g_2(v) = 0$ and $a(v, v) = 0$. Since the kernel of $g_2(\cdot)$ is V_h, and $a(\cdot, \cdot)$ is V_h-elliptic, we get $v = 0$. $\quad\square$

Remark 1.17. *Optimal a priori estimates are available for the variational problems (1.42) and (1.43) by using Lemma 1.16.*

In the rest of this section, we show the relation between the new mortar formulation (1.42) for the special pairing (X_h, M_h^3) and the Dirichlet–Neumann coupling (1.38). It turns out that the non-symmetric product formulation is almost the Dirichlet–Neumann formulation in the case of two subdomains in 2D. Here, we associate the non-mortar side with the subdomain Ω_1. The space X_1 can be directly decomposed into $X_1^I + X_1^\Gamma$. Each element in X_i is extended by zero on the adjacent subdomain, and we will use the same notation for the extended function. Using the decomposition for X_1 and observing that $g_1(v) = g_2(v) = 0$ for $v \in X_1^I$ and $g_2(v) = v$ for $v \in X_1^\Gamma$, (1.42) can be rewritten as a coupled system

$$
\begin{aligned}
a_1(u_h, v) &= (f, v)_{0;\Omega_1}, & v &\in X_1^I \; , \\
b(u_h, g_1(v)) &= 0, & v &\in X_1^\Gamma \; , \\
a(u_h, v - g_2(v)) + b(u_h, g_1(v)) &= (f, v - g_2(v))_0, & v &\in X_2 \; .
\end{aligned}
\tag{1.44}
$$

Comparing (1.38) with (1.44), we find that the first equation is exactly the same. In a second step, we consider the second equation in more detail and obtain for $v = \phi_i$

$$\int_\Gamma (u_{h|\Omega_1} - u_{h|\Omega_2})\psi_i^3 \, d\sigma = u_{h|\Omega_1}(p_i) \int_\Gamma \phi_i \, d\sigma - \int_\Gamma u_{h|\Omega_2}\psi_i^3 \, d\sigma = 0 \; . \tag{1.45}$$

Using the definition of ψ_i^3 and that $\int_\Gamma \phi_i \, d\sigma = 0.5 \, |\text{supp}\, \phi_i|$, we obtain for the interior vertices of Γ

$$u_{h|\Omega_1}(p_i) = \frac{2}{|\text{supp}\, \phi_i|} \int_{\text{supp}\, \phi_i} u_{h|\Omega_2}\psi_i^3 \, d\sigma, \quad 1 \leq i \leq N_\Gamma \; ,$$

where N_Γ is the number of interior vertices on the non-mortar side of Γ. Having (1.37) in mind, (1.45) gives rise to almost the same system as the second equation in (1.38). The only difference is for $i = 1$ and $i = N_\Gamma$; in these cases $\psi_i^3 \neq \psi_{p_i}$. In the general mortar approach, the analysis of the consistency error requires this modification, but only for the vertices on the non-mortar sides sharing one edge with a crosspoint. Finally, we have to compare the equations for $v \in X_2$. Starting with the observation that $v - g_2(v)$ is almost equal to Ev, $v \in X_2$, and using $g_1(v) = 0$ for $v \in X_2$, we find equality except at the two endpoints of Γ of the coupled systems (1.38) and (1.44). Thus the non-symmetric variational problem (1.42) obtained by a local elimination of the Lagrange multiplier is nothing else than a Dirichlet–Neumann coupling of boundary value problems on the different subdomains. On the mortar sides Neumann boundary conditions and on the non-mortar sides Dirichlet boundary conditions are used.

1.3.3 Algebraic Formulations

The introduction of these third mortar formulations was motivated by a need to find new, more efficient, iterative solvers. The positive definite variational formulation (1.14) gives rise to a nonconforming approach where the constrained spaces are non-nested. Working with the equivalent saddle point approach yields an indefinite system for which efficient iterative solvers are often relatively expensive. Two special iterative solvers will be considered in Sect. 2.3 and Sect. 2.4. One is based on the symmetric approach (1.43) and the other one on the unsymmetric Dirichlet–Neumann formulation (1.42). We point out that the use of dual basis functions as Lagrange multiplier space is essential for the efficiency of our new iterative solvers.

In this subsection, we consider the algebraic formulation of the variational problems (1.42) and (1.43) given in Subsect. 1.3.2. We show that the corresponding stiffness matrices can easily be obtained by means of those of the unconstrained product space.

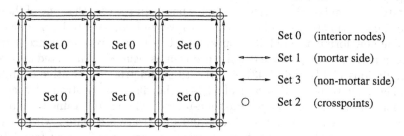

Fig. 1.20. Decomposition of the nodes into sets in 2D

We start with a splitting of the degrees of freedom into different sets; see Fig. 1.20. All nodes in the interior of a subdomain have index zero. The nodes on the interior of the mortar sides have index one. The nodes on the interior of the non-mortar sides have index zero or three; each node corresponding to a basis function of $\widetilde{W}_{0;h_m}(\gamma_m)$ has index three, all other nodes have index zero. In the special case that $\widetilde{W}_{0;h_m}(\gamma_m) = W_{0;h_m}(\gamma_m)$, all nodes on the interior of the non-mortar sides have index three, this is the situation illustrated in Fig. 1.20. All nodes on the boundary of the mortar and non-mortar sides have index two. In 2D, these are the vertices of the triangulations coinciding geometrically with a crosspoint. There are more nodes of this type than crosspoints, since each crosspoint p is associated with n_p nodes, where n_p is the number of subdomains sharing the crosspoint p. In 3D, these are the nodes on the wirebasket of the decomposition, i.e., on $\cup_{m=1}^{M} \partial \gamma_m \setminus \partial \Omega$; see Fig. 1.21. In the right of Fig. 1.21, the situation of four subdomains sharing one edge of the wirebasket is illustrated. The dashed lines symbolize the edges of the triangulations of the different subdomains.

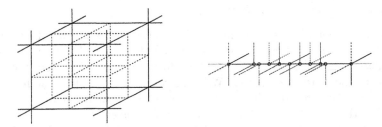

Fig. 1.21. Wirebasket (left) and detail (right)

The variational problems (1.14), (1.15), (1.42), and (1.43) are equivalent, and, in addition, the Lagrange multiplier λ_h can be obtained from u_h by local postprocessing. However, the choice of the underlying mortar formulation will make a big difference for the iterative solution. Here, we only remark that the natural setting for many efficient iterative solvers such as preconditioned conjugate gradient methods or multigrid methods is a symmetric positive definite problem based on nested discrete spaces. In the case that one of these conditions is violated, standard techniques, very often, have to be modified.

The new formulation (1.43) appears to offer a real advantage since it results in a symmetric positive definite problem associated with nested spaces. Formulations (1.14) and (1.43) give rise to symmetric and positive definite problems. However in the case of (1.14), the discrete constrained spaces are nonconforming and, in general, non-nested. The unconstrained product spaces are nested, and (1.15), (1.42), and (1.43) are associated with these spaces. But only (1.43) yields a symmetric and positive definite system. The saddle point formulation (1.15) yields a symmetric but indefinite system, and (1.42) gives rise to a nonsymmetric formulation.

In this subsection, we discuss some algebraic aspects of (1.42) and (1.43). In particular, it is shown how we find (1.42) and (1.43) from (1.15) by local elimination. We consider the corresponding algebraic system, first for the saddle point formulation. Using nodal basis functions and grouping the vectors according to their indices $u_h^T = (u_R^T, u_N^T)$, where $u_R^T := (u_0^T, u_1^T, u_2^T)$ and $u_N := u_3$, (1.15) results in the following indefinite system

$$
\begin{pmatrix} A & B \\ B^T & 0 \end{pmatrix} \begin{pmatrix} u_h \\ \lambda_h \end{pmatrix} := \begin{pmatrix} A_{RR} & A_{RN} & M_R \\ A_{NR} & A_{NN} & D \\ M_R^T & D & 0 \end{pmatrix} \begin{pmatrix} u_R \\ u_N \\ \lambda_h \end{pmatrix} = \begin{pmatrix} f_R \\ f_N \\ 0 \end{pmatrix} =: \begin{pmatrix} f \\ 0 \end{pmatrix} .
$$

$$(1.46)$$

All indices related to the meshsize are suppressed for the stiffness matrices and subvectors. In addition, the same notation is used for functions in the discrete spaces and their coefficient vectors with respect to the nodal basis functions. A_{RR}, A_{RN}, A_{NR} and A_{NN} are the stiffness matrix associated with the bilinear form $a(\cdot, \cdot)$ restricted to the different subsets, and M_R and D are associated with $b(\cdot, \cdot)$. We remark that M_R is a block mass matrix with many zero blocks. Moreover D is a positive definite, diagonal matrix. This is an essential point, and is, as already pointed out, only true if dual basis functions for the Lagrange multipliers are used. Examples of dual basis functions have been discussed in Subsect. 1.2.4.

The non-symmetric formulation (1.42), based on the product space, is obtained by choosing λ_h according to (1.40) and setting $\mu := g_1(v)$ in (1.15). This can be expressed in a very simple way in the algebraic setting

$$
\lambda_h = W^T(f - Au_h) ,
$$
$$
\mu = W^T v ,
$$

where W is defined as

$$
W^T := (0, D^{-1}) .
$$

Then, local elimination of λ_h and μ yields

$$
(\mathrm{Id}, W) \begin{pmatrix} A & B \\ B^T & 0 \end{pmatrix} \begin{pmatrix} \mathrm{Id} \\ -W^T A \end{pmatrix} u_h = (\mathrm{Id} - BW^T)f =: \hat{f} . \qquad (1.47)
$$

It can be easily shown that (1.47) yields $((\mathrm{Id} - BW^T)A + WB^T)u_h = \hat{f}$, and that (1.47) is the algebraic form of (1.42).

A different possibility to obtain a reduced system for u_h is to define μ by

$$
\mu := W^T A(WB^T - \mathrm{Id})v ,
$$

and replace $\lambda_h = W^T(f - Au_h)$ by the equivalent formula

$$
\lambda_h = W^T(f + A(WB^T - \mathrm{Id})u_h) .
$$

We note that the mortar solution satisfies $B^T u_h = 0$. This choice gives rise to a symmetric positive definite system

$$\hat{A}u_h := (\mathrm{Id}, (BW^T - \mathrm{Id})AW) \begin{pmatrix} A & B \\ B^T & 0 \end{pmatrix} \begin{pmatrix} \mathrm{Id} \\ W^T A(WB^T - \mathrm{Id}) \end{pmatrix} u_h = \hat{f} \ .$$

$$(1.48)$$

The following lemma establishes a relation between the algebraic formulation (1.48) and the variational problem (1.43).

Lemma 1.18. *The algebraic formulation of the variational problem (1.43) is given by*

$$\hat{A}u_h = \hat{f} \ .$$

Moreover, the Lagrange multiplier can be obtained, once u_h has been computed, by

$$\lambda_h = W^T(f - Au_h) \ .$$

Proof. To start, we rewrite the linear functional $g_2(v)$ in its algebraic form

$$WB^T v \ .$$

The multiplication with W can be interpreted as a scaled mapping from M_h onto X_h such that the resulting element of X_h is supported in a small strip of width h on the non-mortar side, whereas the multiplication with B^T maps each element in V_h to zero. Using these observations, the algebraic form of variational problem (1.43) can be written as

$$(\mathrm{Id} - BW^T)A(\mathrm{Id} - WB^T)u_h + BW^T AWB^T u_h = (\mathrm{Id} - BW^T)f \ .$$

A straightforward computation completes the proof by comparing this system with (1.48).   □

The matrix $\hat{A}$ can be easily assembled from A, $M := M_R D^{-1}$, and is sparse. Using the indices given at the beginning of this section, it can be written as a 2×2 matrix as

$$\hat{A} = \left(\begin{array}{c|c} A_{RR} + 2M A_{NN} M^T & M A_{NN} \\ -M A_{NR} - A_{RN} M^T & \\ \hline A_{NN} M^T & A_{NN} \end{array} \right) \ . \tag{1.49}$$

1.4 Examples for Special Mortar Finite Element Discretizations

In this subsection, we consider several mortar situations. Each of them provides an interesting insight into the abstract general framework. The flexibility and the wide range of applications is illustrated. Duality arguments play an essential role in our first example. Essential and natural boundary conditions are used to realize the coupling of mixed and standard finite elements. In our second example, we consider the special case that each element is one subdomain, and thus the number of subdomains tends to infinity as the meshsize tends to zero. The settings are given for simplicial triangulations in 2D, but can be generalized without any problem to 3D.

1.4.1 The Coupling of Primal and Dual Finite Elements

In this subsection, we focus on the coupling of two different discretization schemes. We use a mixed finite element discretization on one subdomain and a standard conforming one on the other subdomain. The coupling of mixed finite element discretizations on non-matching grids has been introduced and analyzed in [WY98, AY97, Yot97]. In contrast to standard mortar methods, the finite element trace space of the flux in normal direction at the interface defines a Lagrange multiplier space which does not yield optimal results. In particular, the consistency error is too large compared with the best approximation error. To obtain optimal results, the order of the Lagrange multiplier space has to be increased by one. In the lowest order case, the Lagrange multiplier space has to contain piecewise linear functions and not only the constants.

We observe that the trace space of H^1-functions and the Lagrange multiplier space reflect the duality between $H^{1/2}$ and $H^{-1/2}$ on the interface. In our approach, we use the duality of the roles of Dirichlet and Neumann boundary conditions in the primal and dual setting. Using this duality, we can work on the product space without introducing a Lagrange multiplier. The idea of combining mixed finite element methods with primal ones was originally introduced in [WW98]. Recently, efficient iterative solvers were studied in [LPV99].

For simplicity, we restrict ourselves to the case of two subdomains. Here, $\Omega \subset \mathbb{R}^2$ is decomposed into two nonoverlapping polyhedral subdomains Ω_1 and Ω_2, and we assume that $\text{meas}(\partial\Omega_2 \cap \partial\Omega) > 0$. On Ω_1, we use dual discretization techniques, whereas standard primal approaches are used on Ω_2. The coupling at the interface $\Gamma := \partial\Omega_1 \cap \partial\Omega_2$ is realized without a weak continuity condition between the spaces and without a Lagrange multiplier enforcing the orthogonality of the jump. Instead we take the flux of the solution on Ω_1 to define a Neumann boundary condition on Γ for the boundary value problem on Ω_2, and the solution on Ω_2 as a Dirichlet boundary condition on Γ for the boundary value problem on Ω_1; see Fig. 1.22. We show that proceeding in this way yields a suitable coupling between the two domains. The global discretization error can be bounded by the sum of the local best approximation errors, and optimal a priori bounds can be established.

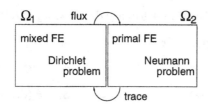

Fig. 1.22. Coupling between dual and primal finite elements

We introduce simplicial triangulations $\mathcal{T}_{h_1}$ and $\mathcal{T}_{h_2}$ on the subdomains Ω_1 and Ω_2 which do not have to match at the interface. The sets of corresponding edges are called $\mathcal{E}_{h_1}$ and $\mathcal{E}_{h_2}$. On Ω_2, where primal techniques are used, we work with standard conforming P_{n_2}-elements, $X_{h_2;n_2} \subset H^1_{0;\Gamma_2}(\Omega_2)$. Here, $H^1_{0;\Gamma_2}(\Omega_2)$ is a subspace of $H^1(\Omega_2)$ satisfying homogeneous Dirichlet boundary conditions on $\Gamma_2 := \partial\Omega \cap \partial\Omega_2$.

The continuous space on Ω_1 is given by $H(\mathrm{div};\Omega_1) \times L^2(\Omega_1)$, where the vector valued Hilbert space $H(\mathrm{div};\Omega_1)$ is defined by $H(\mathrm{div};\Omega_1) := \{\mathbf{q} \in (L^2(\Omega_1))^2 | \ \mathrm{div}\ \mathbf{q} \in L^2(\Omega)\}$. For the discretization, we use Raviart–Thomas finite elements

$$RT_{h_1;n_1} := \{\mathbf{q} \in H(\mathrm{div};\Omega_1) \mid \mathbf{q}_{|_T} \in (P_{n_1}(T))^2 + P_{n_1}(T)\mathbf{x},\ T \in \mathcal{T}_{h_1}\}\ ,$$

of order $n_1 \geq 0$ for the flux; see [RT77], and the space of piecewise polynomials $W_{h_1;n_1} := \{v \in L^2(\Omega_1) \mid v|_T \in P_{n_1}(T),\ T \in \mathcal{T}_{h_1}\}$ of the same order, for the primal variable in Ω_1. We refer to [BF91] for an overview of mixed finite elements.

We remark that no boundary condition has to be imposed on the interface Γ. In contrast to the standard case, Dirichlet boundary conditions are natural boundary conditions for the mixed formulation, i.e., they appear in the definition of the right hand side of the variational problem but are not enforced in the construction of the spaces. For the standard primal approach, the situation is exactly reversed. The Neumann boundary conditions are the natural ones and the Dirichlet boundary conditions are imposed on the space. This duality allows the coupling between Raviart–Thomas and conforming finite elements without Lagrange multiplier on the interface. Using this Dirichlet–Neumann coupling, we find the following discrete variational problem: Find $(\mathbf{j}_{h_1}, u_{h_2}, u_{h_1}) \in RT_{h_1;n_1} \times X_{h_2;n_2} \times W_{h_1;n_1}$ such that

$$
\begin{aligned}
a_1(\mathbf{j}_{h_1}, \mathbf{q}_h) - d(\mathbf{q}_h, u_{h_2}) + b(\mathbf{q}_h, u_{h_1}) &= 0, & \mathbf{q}_h \in RT_{h_1;n_1}\ , \\
d(\mathbf{j}_{h_1}, v_h) + a_2(u_{h_2}, v_h) &= (f, v_h)_{0;\Omega_2}, & v_h \in X_{h_2;n_2}\ , \\
b(\mathbf{j}_{h_1}, w_h) - c(u_{h_1}, w_h) &= -(f, w_h)_{0;\Omega_1}, & w_h \in W_{h_1;n_1}\ .
\end{aligned}
$$
$$(1.50)$$

Here, the bilinear forms $a_i(\cdot,\cdot)$, $i = 1, 2$, $b(\cdot,\cdot)$, $c(\cdot,\cdot)$ and $d(\cdot,\cdot)$ are defined by

$$
\begin{aligned}
a_1(\mathbf{p}_1, \mathbf{q}_1) &:= \int_{\Omega_1} a^{-1}\mathbf{p}_1 \cdot \mathbf{q}_1\ dx, & \mathbf{p}_1, \mathbf{q}_1 \in H(\mathrm{div};\Omega_1)\ , \\
a_2(w_2, v_2) &:= \int_{\Omega_2} (a\nabla v_2\,\nabla w_2 + b\,v_2\,w_2)\,dx, & v_2, w_2 \in H^1(\Omega_2)\ , \\
b(\mathbf{q}_1, v_1) &:= \int_{\Omega_1} \mathrm{div}\mathbf{q}_1\,v_1\,dx, & v_1 \in L^2(\Omega_1),\ \mathbf{q}_1 \in H(\mathrm{div};\Omega_1)\ , \\
c(w_1, v_1) &:= \int_{\Omega_1} b\,w_1\,v_1\,dx, & v_1, w_1 \in L^2(\Omega_1)\ , \\
d(\mathbf{q}_1, v_2) &:= \langle \mathbf{q}_1\mathbf{n}, v_2 \rangle, & \mathbf{q}_1 \in H(\mathrm{div};\Omega_1), v_2 \in H^1(\Omega_2)\ ,
\end{aligned}
$$

where $\langle \cdot, \cdot \rangle$ stands for the duality pairing of $H^{-1/2}(\Gamma)$ and $H^{1/2}(\Gamma)$. The bilinear forms $a_1(\cdot,\cdot)$, $b(\cdot,\cdot)$, and $c(\cdot,\cdot)$ are associated with the dual approach on

Ω_1 whereas $a_2(\cdot, \cdot)$ is a standard $H^1_{0;\Gamma_2}(\Omega_2)$-elliptic bilinear form. The duality of the boundary conditions on Γ is realized by means of $d(\cdot, \cdot)$. Moreover, $d(\cdot, \cdot)$ transfers the boundary conditions between the subdomains. Considering (1.50) in more detail, we find that it has the following saddle point structure

$$\begin{pmatrix} A & B \\ B^T & -C \end{pmatrix} \begin{pmatrix} \mu_h \\ u_{h_1} \end{pmatrix} = \begin{pmatrix} f_2 \\ f_1 \end{pmatrix}, \quad \mu_h := \begin{pmatrix} j_{h_1} \\ u_{h_2} \end{pmatrix},$$

where the operators B and C are associated with the corresponding bilinear forms. The operator A is non symmetric and has the form

$$A := \begin{pmatrix} A_1 & -D \\ D & A_2 \end{pmatrix},$$

where the operators A_1, A_2 and D are associated with the corresponding bilinear forms. The right hand side is given by the linear form $(f, \cdot)_{0;\Omega_i}$, $i = 1, 2$. The kernel of the continuous operator $B_c^T : H(\mathrm{div}; \Omega_1) \times H^1_{0;\Gamma_2}(\Omega_2) \longrightarrow L^2(\Omega_1)$, which is associated with the linear form $b(\cdot, v_1)$, is $\mathrm{Ker} B_c^T := \{(\mathbf{q}_1, v_2) \in H(\mathrm{div}; \Omega_1) \times H^1_{0;\Gamma_2}(\Omega_2) \mid \mathrm{div} \mathbf{q}_1 = 0\}$. It is well known; see, e.g., [BF91], that we have the following equivalence: For $\mathbf{q}_1 \in RT_{h_1;n_1}$, $\mathrm{div} \mathbf{q}_1 = 0$ if and only if

$$b(\mathbf{q}_1, v_1) = 0, \quad v_1 \in W_{h_1;n_1}.$$

Thus the kernel of the discrete operator $B^T : RT_{h_1;n_1} \times X_{h_2;n_2} \longrightarrow W_{h_1;n_1}$ is a subspace of $\mathrm{Ker} B_c^T$.

We further introduce the nonsymmetric bilinear form $a(\sigma, \tau) := a_2(w_2, v_2) + d(\mathbf{p}_1, v_2) + a_1(\mathbf{p}_1, \mathbf{q}_1) - d(\mathbf{q}_1, w_2)$, where $\sigma := (\mathbf{q}_1, v_2)$ and $\tau := (\mathbf{p}_1, w_2)$ are elements of the product space $H(\mathrm{div}; \Omega_1) \times H^1_{0;\Gamma_2}(\Omega_2)$. The norm $\| \cdot \|$ on this product space is inherited by the product topology and is defined by $\|\sigma\|^2 := \|\mathbf{q}_1\|^2_{0;\Omega_1} + \|\mathrm{div} \mathbf{q}_1\|^2_{0;\Omega_1} + \|v_2\|^2_{1;\Omega_2}$. It is now easy to see that the nonsymmetric bilinear form $a(\cdot, \cdot)$ is coercive on $\mathrm{Ker} B_c^T$

$$a(\sigma, \sigma) = a_1(\mathbf{q}_1, \mathbf{q}_1) + a_2(v_2, v_2) \geq c \|\sigma\|^2.$$

Here, we have used that $a_2(\cdot, \cdot)^{1/2}$, restricted to $H^1_{0;\Gamma_2}(\Omega_2)$, is equivalent to the H^1-norm on Ω_2 and that $\mathrm{div} \mathbf{q}_1 = 0$.

An essential tool in establishing a priori bounds for the discretization error is provided by the abstract saddle point theory. A suitable inf-sup condition guarantees that the discretization error can be bounded in terms of the best approximation error. Here, the relevant inf-sup condition is nothing else than the standard one for the mixed finite element scheme on Ω_1

$$\inf_{v_1 \in W_{h_1;n_1}} \sup_{\sigma \in RT_{h_1;n_1} \times X_{h_2;n_2}} \frac{b(\mathbf{q}_1, v_1)}{\|v_1\|_{0;\Omega_1} \|\sigma\|} \geq c.$$

Choosing σ as $(\mathbf{q}_1, 0)$, $\mathbf{q}_1 \in RT_{h_1;n_1}$, this is a standard result for Raviart–Thomas finite elements, and it is satisfied with a constant independent of the meshsize without any further assumptions; see, e.g., [BF91]. In addition, we obtain unique solvability of the saddle point problem (1.50) by taking the continuity of the bilinear forms, the inf-sup condition, and the coercivity of $a(\cdot, \cdot)$ on $\mathrm{Ker}B^T$, into account; see, e.g., [BF91, WW98]. Following [BF91], an optimal a priori bound for the discretization error can be established by means of the best approximation error. Since we are working with unconstrained standard finite element spaces, the approximation properties are well known; see [BF91]. With no matching condition imposed on the discrete spaces at the interface, the analysis as well as the implementation of the method becomes quite simple. A priori estimates of the order of $\min(n_1 + 1, n_2)$ are also obtained

$$\|\mathbf{j} - \mathbf{j}_{h_1}\|_{\mathrm{div};\Omega_1}^2 + \|u - u_{h_1}\|_{0;\Omega_1}^2 + \|u - u_{h_2}\|_{1;\Omega_2}^2$$
$$\leq C \left(h_1^{2(n_1+1)} (|u|_{n_1+1;\Omega_1}^2 + |\mathbf{j}|_{n_1+1;\Omega_1}^2 + |\mathrm{div}\,\mathbf{j}|_{n_1+1;\Omega_1}^2) + h_2^{2n_2} |u|_{n_2+1;\Omega_2}^2 \right)$$
$$(1.51)$$

if the problem has a regular enough solution. Here, $\mathbf{j}$ denotes the continuous flux defined by $\mathbf{j} := a\nabla u \cdot \mathbf{n}$.

Remark 1.19. *The coupling of mixed finite elements on non-matching triangulations has been worked out in [WY98, AY97, Yot97]. As in the standard conforming setting, one has to work with Lagrange multipliers or equivalently a suitable constrained space. In the lowest order case, the Lagrange multiplier space has to be at least piecewise linear, since the optimality of the method is otherwise lost.*

In the next subsection, we show that the saddle point problem (1.50) can be rewritten. Introducing a piecewise constant Lagrange multiplier on the interface, we obtain a mortar finite element method which couples conforming Lagrangian finite elements and nonconforming Crouzeix–Raviart elements. The special characteristic of this approach is that the piecewise constant Lagrange multiplier gives rise to a diagonal mass matrix on the non-mortar sides. In Subsect. 2.3, we discuss how standard multigrid techniques can be applied in such special mortar situations.

1.4.2 An Equivalent Nonconforming Formulation

It is well known that mixed and nonconforming finite element methods are equivalent; see, e.g., [AB85, BF91]. Introducing Lagrange multipliers on the edges of the triangulation, the flux variable as well as the primal variable can be evaluated locally. The resulting Schur complement system is the same as for the positive definite variational problem associated with a nonstandard nonconforming Crouzeix–Raviart discretization; see [BF91]. Furthermore, the mixed finite element solution can be obtained from the nonconforming by local postprocessing. Here, we present a mortar coupling of conforming and nonconforming finite elements which is equivalent to the presented primal dual coupling.

In the rest of this section, we restrict ourselves to the lowest order Raviart–Thomas space, $n_1 = 0$. The dimension of the local flux space is three, and the local space for the discrete primal variable is one dimensional. An equivalence with Crouzeix–Raviart elements can be obtained if we enrich the nonconforming space by local cubic bubble functions. We therefore consider the enriched Crouzeix–Raviart space

$$NC_{h_1} := CR_{h_1} + B_{h_1} ,$$

where CR_{h_1} is the Crouzeix–Raviart space of piecewise linear functions which are continuous at the midpoints of the triangulation $\mathcal{T}_{h_1}$ and equal to zero at the midpoints of any boundary edge $e \in \mathcal{E}_{h_1} \cap \partial\Omega$. B_{h_1} is the space of piecewise cubic bubble functions which vanish on the boundary of the elements.

Restricting the variational problem (1.50), for the moment, to Ω_1 by setting $v_h \in X_{h_2;n_2}$ equal zero, we get

$$\begin{aligned} a_1(\mathbf{j}_{h_1}, \mathbf{q}_h) + b(\mathbf{q}_h, u_{h_1}) &= d(\mathbf{q}_h, u_{h_2}), & \mathbf{q}_h &\in RT_{h_1;0} , \\ b(\mathbf{j}_{h_1}, w_h) - c(u_{h_1}, w_h) &= -(f, w_h)_{0;\Omega_1}, & w_h &\in W_{h_1;0} . \end{aligned} \tag{1.52}$$

If the solution u_{h_2} is known, then the Dirichlet boundary value problem (1.52) for the mixed formulation on Ω_1 can be solved. Comparing (1.50) and (1.52), we find that both variational problems have a saddle point structure. Introducing a Lagrange multiplier and eliminating the variables $\mathbf{j}_{h_1}$ and u_{h_1} in (1.52) yield a symmetric positive definite variational problem. The boundary condition now enters into the definition of the trial space. Let $NC_{g;h_1}$, $g \in L^2(\Gamma)$, be the linear manifold of NC_{h_1} defined by

$$NC_{g;h_1} := \left\{ \psi_h \in NC_{h_1} \mid \int_e \psi_h \, d\sigma = \int_e g \, d\sigma, \, e \in \mathcal{E}_{h_1} \cap \Gamma \right\} .$$

Then, the solution of the saddle point problem (1.52) can be obtained, equivalently, by solving: Find $\Psi_{h_1} \in NC_{u_{h_2};h_1}$ such that

$$a_{NC}(\Psi_{h_1}, \psi_h) = (f, \Pi_0 \psi_h)_{0;\Omega_1}, \quad \psi_h \in NC_{0;h_1} , \tag{1.53}$$

where $a_{NC}(\phi_h, \psi_h) := \sum_{T \in \mathcal{T}_{h_1}} \int_T P_{a^{-1}}(a\nabla\phi_h)\nabla\psi_h + b\Pi_0\phi_h \, \Pi_0\psi_h \, dx$. Here, Π_0 is the L^2-projection onto $W_{h_1;0}$, and $P_{a^{-1}}$ is the weighted L^2-projection, with weight a^{-1}, onto the local Raviart–Thomas space of lowest order which has three degrees of freedom per element; see [AB85, BF91]. We remark that the solution space $NC_{u_{h_2};h_1}$ for the solution in Ω_1 depends on the solution in Ω_2, but that the right hand side in (1.53) does not depend on the solution. Here again, the different roles of the boundary conditions come into play. Using the equivalence of (1.52) and (1.53) in (1.50), we find a global variational problem in Ω: Find $(\Psi_{h_1}, u_{h_2}) \in NC_{u_{h_2};h_1} \times X_{h_2;n_2}$ such that

$$a_{NC}(\Psi_{h_1}, \psi_h) \qquad\qquad = (f, \Pi_0\psi_h)_{0;\Omega_1}, \ \psi_h \in NC_{0;h_1} \ ,$$
$$a_2(u_{h_2}, v_h) + d(P_{a^{-1}}(a\nabla\Psi_{h_1}), v_h) = (f, v_h)_{0;\Omega_2}, \qquad v_h \in X_{h_2;n_2} \ . \tag{1.54}$$

The implementation of (1.54) is based on a different formulation. Considering (1.54), in more detail, it can be seen that it is a minimization problem with constraints, where the constraints depend on the solution. Introducing a Lagrange multiplier space, we can transform (1.54) into a saddle point problem. To do so, we show in a first step that the Dirichlet problem (1.53) can be extended to a variational problem on the whole space NC_{h_1}. Using that $j_{h_1} = P_{a^{-1}}(a\nabla\Psi_{h_1})$ and $u_{h_1} = \Pi_0\Psi_{h_1}$ and applying Green's formula to the second equation of the saddle point problem (1.52), we obtain

$$a_{NC}(\Psi_{h_1}, \psi_h) - d(P_{a^{-1}}(a\nabla\Psi_{h_1}), \psi_h) = (f, \Pi_0\psi_h)_{0;\Omega_1}, \quad \psi_h \in NC_{h_1} \ . \tag{1.55}$$

From the definition of $NC_{g;h_1}$, we find that an element $\chi \in NC_{h_1}$ is in $NC_{g;h_1}$ if and only if

$$\int_\Gamma \mu(\chi - g)\, d\sigma = 0, \quad \mu \in M_{h_1} \ , \tag{1.56}$$

where $M_{h_1} := \{\mu \in L^2(\Gamma) \mid \mu|_e \in P_0(e), \ e \in \mathcal{E}_{h_1} \cap \Gamma\}$. In particular, M_{h_1} is used as a Lagrange multiplier space. The dimension of M_{h_1} is equal to the number of edges in $\mathcal{E}_{h_1} \cap \Gamma$. Finally, we obtain a mortar coupling between conforming and nonconforming finite elements by means of piecewise constant Lagrange multipliers.

Theorem 1.20. *Let $(\Psi_{h_1}, u_{h_2}) \in NC_{u_{h_2};h_1} \times X_{h_2;n_2}$ be the solution of (1.54). Then, $u_M := (\Psi_{h_1}, u_{h_2})$ and $\lambda_M := P_{a^{-1}}(a\nabla\Psi_{h_1})\cdot\mathbf{n}|_\Gamma$ is the unique solution of the following saddle point problem: Find $(u_M, \lambda_M) \in (NC_{h_1} \times X_{h_2;n_2}) \times M_{h_1}$ such that*

$$a(u_M, v) - \hat{d}(\lambda_M, v) = f(v), \quad v \in NC_{h_1} \times X_{h_2;n_2} \ ,$$
$$\hat{d}(\mu, u_M) \qquad\qquad = 0, \qquad \mu \in M_{h_1} \ . \tag{1.57}$$

Here, the bilinear and linear forms are given by, $v := (v_1, v_2)$, $w := (w_1, w_2)$:

$$a(w, v) := a_2(w_2, v_2) + a_{NC}(w_1, v_1), \ v, w \in NC_{h_1} \times X_{h_2;n_2} \ ,$$
$$\hat{d}(\mu, v) := \int_\Gamma \mu(v_1 - v_2)\, d\sigma, \qquad\qquad \mu \in M_{h_1} \ ,$$
$$f(v) := (f, v_2)_{0;\Omega_2} + (f, \Pi_0 v_1)_{0;\Omega_1} \ .$$

Proof. The assertion is an easy consequence of (1.54) by using (1.55) and (1.56). $\square$

Theorem 1.20 states the equivalence of (1.50) and (1.57) in the case $n_1 = 0$. We obtain the solution of (1.50) by a local postprocessing from the solution of (1.57), from the formulas $j_{h_1} = P_{a^{-1}}(a\nabla u_M|_{\Omega_1})$, $u_{h_1} = \Pi_0 u_M|_{\Omega_1}$ and $u_{h_2} = u_M|_{\Omega_2}$. The a priori bound (1.51) and Theorem 1.20 guarantee that the discretization error of $u - u_M$ in the energy norm is of order h. Comparing (1.57) with (1.15), we find that (1.57) represents the saddle point formulation of a mortar coupling.

We recall that the saddle point problem (1.50) realizes the coupling of dual and primal finite elements methods. Using the equivalent nonconforming approach instead of the mixed method, we arrive at a mortar finite element method expressing the coupling of conforming and nonconforming spaces. The analysis of the resulting discrete problem could be also done within the abstract mortar framework. The piecewise constant Lagrange multiplier $\lambda_M = j_{h_1} \cdot n|_\Gamma$ is associated with the side of the nonconforming discretization, and the mass matrix on the non-mortar side is diagonal. We find that the mortar side is associated with the conforming discretization and the non-mortar side with the nonconforming discretization. Figure 1.23 illustrates the relation between the different couplings.

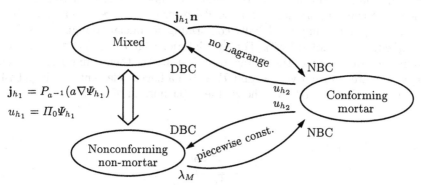

Fig. 1.23. Equivalence between primal dual coupling and mortar coupling

Remark 1.21. *For the implementation, we will eliminate the cubic bubble functions in (1.57) locally; see, e.g., [AB85, BF91]. We then obtain the standard variational problem for Crouzeix–Raviart elements, where the right hand side f is replaced by $\Pi_0 f$ in the special case of $b = 0$ and a piecewise constant diffusion coefficient a. The solution Ψ_{h_1} of the nonconforming problem is given by*

$$\Psi_{h_1}|_T = u_{h_1}|_T + \frac{5}{12} \sum_{i=1}^{3} h_{e_i}^2 \, \Pi_0 f|_T (\lambda_1 \lambda_2 \lambda_3), \quad T \in \mathcal{T}_{h_1},$$

where λ_i, $1 \leq i \leq 3$, are the barycentric coordinate functions, and h_{e_i} is the length of the edge $e_i \subset \partial T$, $1 \leq i \leq 3$. Here, u_{h_1} stands for the Crouzeix–Raviart part of the mortar finite element solution of (1.57).

1.4.3 Crouzeix–Raviart Finite Elements

A large class of nonconforming methods can be analyzed within the mortar framework, here we present the well known Crouzeix–Raviart elements in the context of mortar finite elements; see [CR74]. Reformulating standard nonconforming finite elements on matching triangulations as mortar finite elements can also be carried over to rotated bilinear elements on quadrilateral triangulations and to 3D. Previously, we have decomposed the domain Ω into a fixed number of subdomains Ω_k, and we have implicitly assumed that the number K of subdomains is small compared to the number of elements N_h of the global triangulation $\mathcal{T}_h$. In particular, within a multilevel approach the ratio between the number of subdomains and elements K/N_h tends to zero with $h \to 0$. However, neither the number of subdomains nor the ratio K/N_h enter into the constants of the a priori bounds and the inf-sup condition.

To rewrite the Crouzeix–Raviart elements as a mortar method, we have to consider the extreme case that the decomposition of Ω is given by the finite element triangulation and that the number of subdomains tends to infinity as the discretization parameter of $\mathcal{T}_h$ tends to zero. Here, we assume that $\mathcal{T}_h$ is a shape regular family of globally conforming simplicial triangulations. The set of edges is denoted by $\mathcal{E}_h$. For each interior edge $e = \partial T_i \cap \partial T_o$, we have to fix the orientation of the normal direction $\mathbf{n}_e$. We then define T_o such that $\mathbf{n}_e$ is its outer normal vector, and T_i as the adjacent element; see Fig. 1.24. In the case that $e \subset \partial\Omega$, we choose the outer normal of $\partial\Omega$ as $\mathbf{n}_e$.

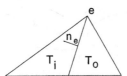

Fig. 1.24. Orientation of the normal vector $\mathbf{n}_e$

Starting with the following decomposition of Ω

$$\overline{\Omega} = \bigcup_{T \in \mathcal{T}_h} \overline{T} \ ,$$

and proceeding as in Sect. 1.1, we find that the interior edges of the triangulation are the interfaces. We also include the boundary edges of the triangulation into the set of interfaces. On each subdomain T, we choose the discrete space of linear functions $P_1(T)$ and define the unconstrained product space by

$$X_h := \prod_{T \in \mathcal{T}_h} P_1(T) \ . \tag{1.58}$$

In Sect. 1.1, we have seen that the bilinear form $a(\cdot, \cdot)$

$$a(v, w) := \sum_{T \in \mathcal{T}_h} \int_T a \nabla v \cdot \nabla w + b \, v \, w \, dx, \quad v, w \in \prod_{T \in \mathcal{T}_h} H^1(T) \ ,$$

is uniform elliptic on a suitable subspace of $\prod_{T \in \mathcal{T}_h} H^1(T) \times \prod_{T \in \mathcal{T}_h} H^1(T)$. To have ellipticity it is sufficient to guarantee that the mean value of the jumps vanishes across the edges. The nonconforming space V_h is thus defined as a subspace of X_h satisfying matching conditions at the interfaces

$$V_h := \left\{ v \in X_h \mid \int_e [v] \, d\sigma = 0, \ e \in \mathcal{E}_h \right\} \ . \tag{1.59}$$

Here, $[v] := v_{|T_o} - v_{|T_i}$ for an inner edge, while for an edge $e \subset \partial \Omega$, it is defined as the trace. From the definition of V_h, it follows that $V_h \subset Y$ and thus the bilinear form $a(\cdot, \cdot)$ is uniformly elliptic on $V_h \times V_h$, which guarantees the unique solvability of the symmetric variational problem: Find $u_h \in V_h$ such that

$$a(u_h, v) = (f, v)_0, \quad v \in V_h \ . \tag{1.60}$$

The definition (1.59) is equivalent to (1.7) with the special local Lagrange multiplier space

$$M(e) := P_0(e) \ ,$$

and the global one is given by $M_h := \prod_{e \in \mathcal{E}_h} M(e)$.

We have modified the definition of V_h in one respect, in comparison to Sect. 1.1. The boundary of Ω is now part of the union of the interfaces and thus, the homogeneous Dirichlet boundary condition is satisfied only in the weak form

$$\int_e v \, d\sigma = 0, \quad e \in \partial \Omega, v \in V_h \ .$$

The constraints at the interface for $v \in V_h$

$$\int_e [v] \, d\sigma = 0, \quad e \in \mathcal{E}_h \ ,$$

force a piecewise linear function to be continuous at the midpoint m_e of an interior edge e and to equal zero at the midpoint of a boundary edge $e \subset \partial \Omega$.

Recalling the definition of Crouzeix–Raviart elements of lowest order

$$CR_h := \left\{ v \in L^2(\Omega) \mid v_{|T} \in P_1(T), T \in \mathcal{T}_h, v(m_e)_{|T_o} = 0, e \in \mathcal{E}_h, e \subset \partial \Omega \ , \right.$$
$$\left. v_{|T_i}(m_e) = v_{|T_o}(m_e), e \in \mathcal{E}_h, e \subset \Omega \right\} \ ,$$

we find that $V_h = CR_h$. Thus, the positive definite nonconforming variational problem for the Crouzeix–Raviart elements is exactly the same as the nonconforming formulation of the mortar method associated with the special decomposition (1.58) and (1.59). In contrast to the more general case of Sect. 1.1, the nodal basis functions of V_h have local support in exactly two elements.

In the rest of this subsection, we introduce and analyze a simplified a posteriori error estimator. The starting point for the construction of this error estimator is the corresponding saddle point problem: Find $(u_h, \lambda_h) \in X_h \times M_h$ such that

$$
\begin{aligned}
a(u_h, v) + b(v, \lambda_h) &= (f, v)_0, & v \in X_h \ , \\
b(u_h, \mu) &= 0, & \mu \in M_h \ ,
\end{aligned}
\tag{1.61}
$$

where $b(w, \mu) := \sum_{e \in \mathcal{E}_h} \int_e [w] \mu \, d\sigma$. Observing that the global Lagrange multiplier space M_h is nothing but the product space of one dimensional spaces, the unique solvability can be obtained without explicitly considering an inf-sup condition. Let $u_h \in V_h$ be the solution of (1.60). Then, the Lagrange multiplier is given locally by

$$
\lambda_h|_e := (f, v_{e;T_o})_0 - a(u_h, v_{e;T_o}) = -(f, v_{e;T_i})_0 + a(u_h, v_{e;T_i}) \ ,
\tag{1.62}
$$

where $v_{e;T_i}$ and $v_{e;T_o} \in X_h$ are local basis functions satisfying $\int_{\hat{e}} v_{e;T_i} \, d\sigma = \int_{\hat{e}} v_{e;T_o} \, d\sigma = \delta_{e\hat{e}}$, supp $v_{e;T_i} = \overline{T}_i$ and supp $v_{e;T_o} = \overline{T}_o$. It can be easily verified that the Lagrange multiplier is well defined and that (u_h, λ_h) satisfies (1.61).

Reliable and efficient a posteriori error estimators are considered in [CJ97, CJ98, DDPV96, Woh99c] for Crouzeix–Raviart elements. In contrast to standard conforming approaches, local a posteriori error estimators have to include an additional term measuring the discontinuity of the finite element solution. Basically, the definitions of the local contributions are the same and can be characterized by

$$
\eta_T^2 := \sum_{e \subset \partial T} \left(\frac{\omega_e}{h_e} \|[u_h]\|_{0;e}^2 + h_e \tilde{\omega}_e \|[\frac{a \partial u_h}{\partial \mathbf{n}_e}]\|_{0;e}^2 \right) + h_T^2 \|\Pi_1 f - b u_h\|_{0;T}^2 \ ,
\tag{1.63}
$$

where $\omega_e, \tilde{\omega}_e > 0$ are suitable weighting factors which do not depend on the meshsize but, in general, on the coefficients. The L^2-projection Π_1 onto piecewise linear functions can be replaced by the one onto piecewise constants. The difference between them can be bounded by $h_T^2 \|f - \Pi_0 f\|_{0;T}^2$ which is a higher order term compared with the other components of the error estimator, and it can therefore be neglected.

We can now show that a simplified a posteriori error estimator can be obtained. The starting point of the construction, is the explicit representation of the Lagrange multiplier given in (1.62). This simplified residual based error estimator for Crouzeix–Raviart finite elements was first introduced in [Woh99c]. By considering the saddle point problem (1.61), it can be shown that the term $\||[\frac{a\partial u_h}{\partial \mathbf{n}_e}]\||_{0;e}$ is redundant. Applying Green's formula on (1.62), the Lagrange multiplier can be obtained in terms of $a\partial u_h/\partial \mathbf{n}_e$ and the elementwise residual. The explicit form of the Lagrange multiplier yields

$$[\frac{a\partial u_h}{\partial \mathbf{n}_e}]|_e = \int\limits_{T_i} (\Pi_1 f - bu_h)v_{e;T_i}\, dx + \int\limits_{T_o} (\Pi_1 f - bu_h)v_{e;T_o}\, dx\ .$$

Since $[a\frac{\partial u_h}{\partial \mathbf{n}_e}]$ is constant on each edge, we find an upper bound for the second term in (1.63)

$$h_e\||[\frac{a\partial u_h}{\partial \mathbf{n}_e}]\||_{0;e}^2 \leq \frac{1}{3}|T_i \cup T_o|\,\|\Pi_1 f - bu_h\|_{0;T_i \cup T_o}^2\ . \tag{1.64}$$

These preliminary considerations motivate the introduction of our simplified error estimator. The local contributions $\hat{\eta}_T$ are defined by

$$\hat{\eta}_T^2 := \sum_{e \subset \partial T} \frac{\omega_e}{h_e}\||[u_h]\||_{0;e}^2 + h_T^2\|\Pi_1 f - bu_h\|_{0;T}^2\ .$$

The following lemma shows a quasi-local equivalence between the two a posteriori error estimators.

Lemma 1.22. *There exists a constant such that*

$$\hat{\eta}_T \leq \eta_T \leq C\,(\hat{\eta}_T + \sum_{\substack{e \subset \partial T \\ \partial T_e \cap \partial T = e}} \hat{\eta}_{T_e})\ .$$

Proof. The proof is an easy consequence of the definition of the local contribution of the error estimators and (1.64).   □

The evaluation of the local contributions $\hat{\eta}_T$ within an adaptive multilevel scheme is extremely simple. Only one contribution per element and one per edge have to be computed. In particular, no linear system of equations has to be solved. Numerical results for this simplified error estimator can be found in the next section.

1.5 Numerical Results

In this section, we collect and describe the results from different series of numerical experiments with several mortar settings in 2D. The methods are tested on different types of examples, and the numerical results confirm the theory. Throughout this work, the implementation is based on the finite element toolbox ug, [BBJ+97].

To show the advantages of mortar methods, we present some examples which illustrate the flexibility and efficiency of these domain decomposition techniques. A rotating geometry, subdomains with reentrant corners, and a region with an extremely bad aspect ratio are discussed. These examples are borrowed from [WW99]. For more details and further examples, we refer to [WW98, WW99].

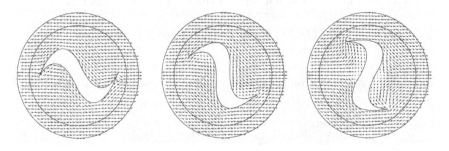

Fig. 1.25. The flux of the solution on a rotating geometry

We start with a time-dependent problem. For standard conforming discretization schemes, remeshing is very often required after each time-step. In particular in 3D, this can be very expensive and inefficient. Using mortar techniques, frequent remeshing can be avoided, even in the case of a rotating geometry and for an arbitrary ratio between the meshsize and the time-step. In this example, the domain Ω is not simply connected, and is decomposed into two subdomains, where the interior one rotates with a fixed angular speed; see Fig. 1.25. The interface between the interior and the outer subdomain is a circle, and we use discontinuous piecewise constant Lagrange multipliers. At the outer boundary, homogeneous Dirichlet conditions are given whereas at the inner homogeneous Neumann condition are assumed. The elliptic problem under consideration is the Poisson problem with a constant right hand side.

The second example shows an application of the coupling of the primal and dual techniques as presented in Sect. 1.4.1. The implementation is based on the equivalent nonconforming approach given in Subsect. 1.4.2. We consider a simple flow problem in heterogeneous media, modeled by Darcy's law for a pressure potential u and the flow $a\nabla u$. The domain is decomposed into several subdomains, where Ω_1 represents a simply connected polygonal channel region; see Fig. 1.26. On Ω_1, we use nonconforming finite elements; the other subdomains are discretized by standard conforming finite elements of lowest order. We choose the diffusion parameters $a_1 := 0.001$ in Ω_1, and $a_2 := 1$ elsewhere. Inflow and outflow boundary conditions for the entrance and exit of the channel region are used and homogeneous Neumann boundary conditions elsewhere. On Ω_1 a finer mesh is used, and the triangulations do not match at the interfaces. This example has been inspired by the cover figure of [BS94].

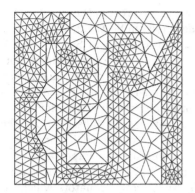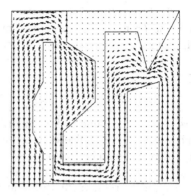

Fig. 1.26. Initial mesh for the channel domain (left) and flux (right)

In the next example, we apply the mortar technique to a linear elasticity problem. The Laplace operator is replaced by the following variational problem: Find $u_h \in V_h$ such that

$$\sum_{k=1}^{K} \int_{\Omega_k} (2\mu_k\, \varepsilon(u_h) : \varepsilon(v) + \lambda_k \operatorname{div}(u_h) \operatorname{div}(v))\, dx = (f, v)_0, \quad v \in V_h \ ,$$

where the Lamé constants μ_k, λ_k depend on the subdomains. The usual notations are used, and V_h is a suitable constrained space in the sense of mortar techniques. We consider a composite material constructed from large bricks of hard steel Ω_i, $2 \le i \le K$, joined with thin layers of a less hard material Ω_1; see Fig. 1.27. The following parameters are used: $\lambda_k = 110743$, $\mu_k = 80193$, $2 \le k \le K$, and $\lambda_1 = 135671$, $\mu_1 = 67837$. Our choice of this geometry has been inspired by the picture of bricks and mortar often used in introductions to mortar finite element methods.

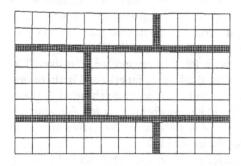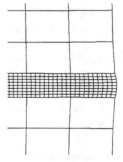

Fig. 1.27. Deformation of the composite (left) and zoom of a thin layer (right)

In Subsect. 1.5.1, we consider the discretization errors in the energy norm and the L^2-norm for the mortar methods described in Sect. 1.2. We compare the four different Lagrange multiplier spaces introduced in Subsect. 1.2.4. Subsection 1.5.2 concerns the numerical performance of a non-optimal mortar method, for which the best approximation property of the constrained space is violated. The a priori estimates in the energy norm are only of order $h^{1/2}$. Nevertheless, we find that asymptotically the discretization errors are almost the same as for the other mortar methods. In Subsect. 1.5.3, the influence of the choice of the non-mortar side is considered. Uniform and adaptive refinement techniques are used to illustrate the importance of a suitable choice of the mortar side in the case of discontinuous coefficients and uniform refinement techniques. A posteriori error estimators which include the weighted L^2-norm of the jump on the non-mortar sides generate for both situations optimal triangulations. The adaptive refinement in the neighborhood of the interface depends strongly on the choice of the mortar sides. Finally in Subsect. 1.5.4, we study the influence of the jump in the coefficients on the adaptive refinement process.

An essential tool for any efficient numerical solution process is also a good iterative solver. In this section, we do not discuss and analyze these matters, but note that different efficient solvers are developed in Chap. 2. Here, we remark only that well known methods like multigrid techniques, iterative substructuring methods, and hierarchical basis preconditioners have been extended to the mortar setting. In the experiments reported here, we have used a multigrid method for saddle point problems as our iterative solver. This approach has been applied in [WW98, WW99], and further analyzed in [BD98, BDW99, Bra01].

1.5.1 Influence of the Lagrange Multiplier Spaces

We now present some numerical results in 2D illustrating the discretization errors for different mortar settings. We recall that in the standard mortar approach the Lagrange multiplier space M_h^1 is used, whereas alternative Lagrange multiplier spaces M_h^l, $2 \leq l \leq 4$, are proposed in Subsect. 1.2.4. Using the same notations as before, u_h^i denotes the mortar finite element solution associated with the Lagrange multiplier space M_h^i. We recall that the unconstrained product space is the same for all four mortar solutions. The definitions of M_h^3 and M_h^4 are based on dual basis functions. Furthermore, the elements in M_h^1 and M_h^3 are piecewise linear, whereas those in M_h^2 and M_h^4 are piecewise constant. Piecewise linear conforming finite elements are used on each subdomain. A comparison of the discretization errors of u_h^1 and u_h^3 can also be found in [Woh00a] for Examples 1–3.

In this subsection, we use only uniform refinement. Starting with an initial triangulation $\mathcal{T}_0$, the triangulation $\mathcal{T}_l$ on level l is obtained by uniform refinement of $\mathcal{T}_{l-1}$; each element of $\mathcal{T}_{l-1}$ is decomposed into four congruent subelements.

The discretizations discussed in Sect. 1.2 are compared for the following four examples. Example 1 is given by: $-\Delta u = f$ on $(0,1)^2$, where the right hand side f and the Dirichlet boundary conditions are chosen so that the exact solution is $(\exp(-500s)-1)(\exp(-500t)-1)(\exp(-500yy)-1)(1-3r)^2$. Here, $s := (x-1/3)^2$, $t := (x-2/3)^2$, $xx := (x-1/2)^2$, $yy := (y-1/2)^2$ and $r := xx + yy$. The isolines of the solution and the initial triangulation are given in Fig. 1.28.

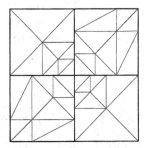
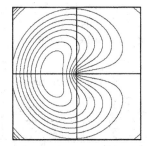

Fig. 1.28. Decomposition into 9 subdomains and initial triangulation (left) and isolines of the solution (right), (Example 1)

The domain is decomposed into nine subdomains, defined by $\Omega_{ij} := ((i-1)/3, i/3) \times ((j-1)/3, j/3)$, $1 \le i, j \le 3$, and the triangulations do not match at the interfaces. We observe two different situations at the interfaces; the isolines of the solution are almost parallel to $\partial\Omega_{11} \cap \partial\Omega_{12}$ whereas at $\partial\Omega_{11} \cap \partial\Omega_{21}$ the angle between the isolines and the interface is bounded away from zero on a large part of the interface. Where the isolines are orthogonal on the interface, the flux vanishes. We recall that the discrete Lagrange multiplier is an approximation of the flux.

Table 1.1. Discretization errors in the L^2-norm, (Example 1)

level	# elem.	$\|u - u_h^1\|_0$	$\|u - u_h^3\|_0$	$\|u - u_h^2\|_0$	$\|u - u_h^4\|_0$
0	72	2.021163e-0	2.021306e-0	2.021196e-0	2.021299e-0
1	288	1.017372e-1	1.014502e-1	1.014460e-1	1.013067e-1
2	1152	1.166495e-1	1.166435e-1	1.166459e-1	1.166458e-1
3	4608	9.482530e-3	9.476176e-3	9.478390e-3	9.476248e-3
4	18432	2.802710e-3	2.797809e-3	2.800444e-3	2.797812e-3
5	73728	7.130523e-4	7.121334e-4	7.126101e-4	7.121334e-4
6	294912	1.789436e-4	1.788082e-4	1.788774e-4	1.788080e-4

In Table 1.1 and Table 1.2, the discretization errors are given in the L^2-norm and in the energy norm, respectively. The columns are ordered in the following way: Columns 3 and 4 show the results for the mortar solutions where the Lagrange multiplier spaces are based on piecewise linear functions. In Columns 5 and 6, we find the results in the case of piecewise constant Lagrange multiplier spaces. Furthermore Columns 4 and 6 correspond to the choice of a biorthogonal basis. The observed asymptotic rates confirm the theory. We find that the energy error is of order h whereas the error in the L^2-norm is of order h^2. There is no significant difference in the accuracy between the different mortar algorithms on any level neither in the L^2-norm nor in the energy norm. On level 6, the difference in the accuracy between the best u_h^4 and worst u_h^1 mortar solution is less than 0.08% in the L^2-norm and less than 0.04% in the energy norm. The influence of the choice of the Lagrange multiplier space on the accuracy of the solution is negligible.

Table 1.2. Discretization errors in the energy norm, (Example 1)

level	# elem.	$\|u - u_h^1\|$	$\|u - u_h^3\|$	$\|u - u_h^2\|$	$\|u - u_h^4\|$
0	72	1.147900e+1	1.147984e+1	1.147918e+1	1.147980e+1
1	288	3.042101e-0	3.034778e-0	3.036952e-0	3.033598e-0
2	1152	1.945246e-0	1.946163e-0	1.945328e-0	1.946169e-0
3	4608	1.114075e-0	1.113506e-0	1.113775e-0	1.113507e-0
4	18432	5.928275e-1	5.923121e-1	5.925919e-1	5.923119e-1
5	73728	2.981975e-1	2.980159e-1	2.981087e-1	2.980157e-1
6	284912	1.492382e-1	1.491841e-1	1.492114e-1	1.491839e-1

In our second example, we consider the unit square with a slit, $\Omega :=(0,1)^2 \setminus [0.5, 1) \times \{0.5\}$, decomposed into four subdomains; see Fig. 1.29. Here, we have no H^2-regularity and a $\mathcal{O}(h)$ and $\mathcal{O}(h^2)$ behavior of the discretization errors in the energy norm and L^2-norm, respectively, cannot be expected. The right hand side f and the Dirichlet boundary conditions of $-\Delta u = f$ are chosen such that the exact solution is given by $(1-3r^2)^2 r^{1/2} \sin(1/2\phi)$, where $x - 1/2 = r \cos \phi$, and $y - 1/2 = r \sin \phi$. The solution has a singularity in the center of the domain.

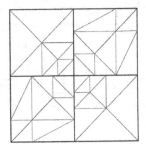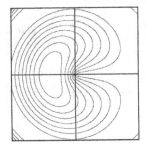

Fig. 1.29. Decomposition into 4 subdomains and initial triangulation (left) and isolines of the solution (right), (Example 2)

The discretization errors are given in Tables 1.3 and 1.4. In this case, we observe a significant difference in the performance of the different mortar methods. The discretization errors in the L^2-norm of the alternative mortar methods with M_h^3 and M_h^4 as Lagrange multiplier spaces are asymptotically better than those of the others. It seems that the biorthogonal Lagrange multiplier spaces M_h^3 and M_h^4 provide a better approximation in this special case without full regularity. We note that continuity of the flux is only guaranteed if the weak solution is in $H^{2+\epsilon}(\Omega)$, $\epsilon > 0$. As in Example 1, the discretization errors measured in the energy norm are comparable. The standard mortar approach u_h^1 and u_h^2 give slightly better results than u_h^3 and u_h^4.

Table 1.3. Discretization errors in the L^2-norm, (Example 2)

level	# elem.	$\|u - u_h^1\|_0$	$\|u - u_h^3\|_0$	$\|u - u_h^2\|_0$	$\|u - u_h^4\|_0$
0	44	4.896283e-2	4.861265e-2	4.882552e-2	4.861265e-2
1	176	1.651238e-2	1.619017e-2	1.637565e-2	1.619057e-2
2	704	4.488552e-3	4.281367e-3	4.394189e-3	4.281791e-3
3	2816	1.254716e-3	1.125460e-3	1.190041e-3	1.125644e-3
4	11264	3.878438e-4	3.046049e-4	3.432019e-4	3.047055e-4
5	45056	1.401538e-4	8.680669e-5	1.109596e-4	8.686759e-5
6	180224	5.883500e-5	2.649174e-5	4.153425e-5	2.653024e-5

Table 1.4. Discretization errors in the energy norm, (Example 2)

level	# elem.	$\|u - u_h^1\|$	$\|u - u_h^3\|$	$\|u - u_h^2\|$	$\|u - u_h^4\|$
0	44	6.000955e-1	6.050778e-1	6.015198e-1	6.050778e-1
1	176	3.553279e-1	3.584246e-1	3.563046e-1	3.584008e-1
2	704	2.045833e-1	2.069586e-1	2.053318e-1	2.069517e-1
3	2816	1.232939e-1	1.252113e-1	1.238978e-1	1.252059e-1
4	11264	7.824813e-2	7.975380e-2	7.872322e-2	7.974960e-2
5	45056	5.184650e-2	5.298379e-2	5.220587e-2	5.298063e-2
6	180224	3.536026e-2	3.619496e-2	3.562424e-2	3.619266e-2

Our next example illustrates the influence of discontinuous coefficients. We consider the diffusion equation $-\operatorname{div} a\nabla u = f$, on $(0,1)^2$, where the coefficient a is discontinuous. The unit square Ω is decomposed into four non overlapping subdomains $\Omega_{ij} := ((i - 1)/2, i/2) \times ((j - 1)/2, j/2)$, $i, j \in \{1, 2\}$, as in Fig. 1.30. The coefficients on the subdomains are given by $a_{11} = a_{22} = 0.00025$, $a_{12} = a_{21} = 1$. The right hand side f and the Dirichlet boundary conditions are chosen to match a given exact solution, $(x - 0.5)(y - 0.5)\exp(-10((x - 0.5)^2 + (y - 0.5)^2))/a$. This solution is continuous with value zero at the interfaces. Furthermore, the jump of the flux, $[a\nabla u \cdot \mathbf{n}]$, vanishes on the interfaces. Because of the discontinuity of the coefficients, we use a highly non-matching triangulation at the interface; see Fig. 1.30.

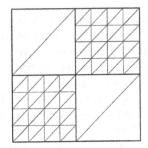 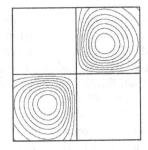

Fig. 1.30. Decomposition into 4 subdomains and initial triangulation (left) and isolines of the solution (right), (Example 3)

The discretization errors in the L^2-norm and in the energy norm are given for the different mortar discretizations in Table 1.5 and Table 1.6, respectively. We observe an $\mathcal{O}(h^2)$ behavior for the discretization errors in the L^2-norm, and that the energy errors are of order h.

Table 1.5. Discretization errors in the L^2-norm, (Example 3)

level	# elem.	$\|u - u_h^1\|_0$	$\|u - u_h^3\|_0$	$\|u - u_h^2\|_0$	$\|u - u_h^4\|_0$
0	68	3.184810e-0	2.981474e-0	3.113584e-0	2.981474e-0
1	272	9.416096e-1	9.358117e-1	9.399402e-1	9.358117e-1
2	1088	2.425569e-1	2.431694e-1	2.427259e-1	2.431694e-1
3	4352	6.093936e-2	6.103994e-2	6.096660e-2	6.103994e-2
4	17408	1.524479e-2	1.525489e-2	1.524751e-2	1.525489e-3
5	69632	3.811271e-3	3.812137e-3	3.811503e-3	3.812137e-3
6	278528	9.527881e-4	9.528569e-4	9.528064e-4	9.528568e-4

Table 1.6. Discretization errors in the energy norm, (Example 3)

level	# elem.	$\|u - u_h^1\|$	$\|u - u_h^3\|$	$\|u - u_h^2\|$	$\|u - u_h^4\|$
0	68	1.173889e-0	1.199259e-0	1.181552e-0	1.199259e-1
1	272	6.115732e-1	6.187439e-1	6.133283e-1	6.187439e-1
2	1088	3.083728e-1	3.094938e-1	3.086346e-1	3.094938e-1
3	4352	1.545031e-1	1.546515e-1	1.545374e-1	1.546515e-1
4	17408	7.729229e-2	7.731113e-2	7.729663e-2	7.731113e-2
5	69632	3.865144e-2	3.865380e-2	3.865198e-2	3.865380e-2
6	278528	1.932641e-2	1.932670e-2	1.932648e-2	1.932670e-2

As in Example 1, there is only a minimal difference in the performance of the mortar approaches with conforming P_1-elements on the subdomains. Again the numerical results confirm the theory. On level 6, the difference in the accuracy between the best u_h^1 and worst u_h^4 mortar solution is less than 0.008% in the L^2-norm and less than 0.002% in the energy norm.

In the last example in this subsection, we combine discontinuous coefficients and a weak solution with a singularity. The unit square is decomposed into two subdomains. Ω_1 is a L-shape domain and $\Omega_2 := \Omega \setminus \overline{\Omega}_1$; see Fig. 1.31. We have no H^2-regularity, and no $\mathcal{O}(h)$ and $\mathcal{O}(h^2)$ behavior of the discretization errors in the energy norm and in the L^2-norm, respectively, can be expected. The Dirichlet boundary conditions of $-\mathrm{div}\,a\nabla u = 0$ are chosen so that the exact solution is given by $r^\alpha \sin(\alpha\phi + \phi_1)$ on Ω_1 and by $\beta r^\alpha \sin(\alpha\phi + \phi_2)$ on Ω_2. Here, the parameters are $x - 1/2 = r\cos\phi$, and $y - 1/2 = r\sin\phi$, $\alpha := 0.6675$, $\phi_1 := (1 - 0.75\alpha)\pi$, $\phi_2 := (2 - 1.75\alpha)\pi$, $\beta := \sin(\phi_1)/\sin(2\pi\alpha + \phi_2)$, $a_{|\Omega_1} := 1$ and $a_{|\Omega_2} := -\tan(0.25\alpha\pi)/\tan(0.75\alpha\pi)$. The solution is continuous and non zero at the interface, and the normal derivative, but not the flux, has a jump at the interface.

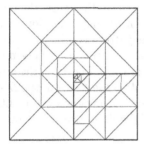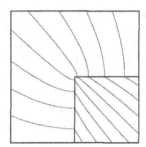

Fig. 1.31. Decomposition into 2 subdomains and initial triangulation (left) and isolines of the solution (right), (Example 4)

Tables 1.7 and 1.8 show the discretization errors for the different Lagrange multiplier spaces. As before in Examples 1 and 3, the influence of the Lagrange multiplier space on the discretization errors is negligible. From the beginning, the errors in the energy norm are almost the same in all four cases. The L^2-norm is more sensitive to the choice of the Lagrange multiplier space. But even for this norm, the difference in the accuracy is smaller than 0.1% after two refinement steps.

Table 1.7. Discretization errors in the L^2-norm, (Example 4)

level	# elem.	$\|u - u_h^1\|_0$	$\|u - u_h^3\|_0$	$\|u - u_h^2\|_0$	$\|u - u_h^4\|_0$
0	60	6.113544e-3	6.154868e-3	6.124583e-3	6.154868e-3
1	240	2.402816e-3	2.416258e-3	2.406854e-3	2.416258e-3
2	960	9.606488e-4	9.610276e-4	9.607471e-4	9.608221e-4
3	3840	3.851735e-4	3.851173e-4	3.851625e-4	3.850316e-4
4	15360	1.546044e-4	1.545783e-4	1.545983e-4	1.545459e-4
5	61440	6.194618e-5	6.193827e-5	6.194430e-5	6.192830e-5
6	245760	2.475497e-5	2.475282e-5	2.475445e-5	2.475004e-5

Table 1.8. Discretization errors in the energy norm, (Example 4)

level	# elem.	$\|u - u_h^1\|$	$\|u - u_h^3\|$	$\|u - u_h^2\|$	$\|u - u_h^4\|$
0	60	1.410069e-1	1.410099e-1	1.410075e-1	1.410099e-1
1	240	8.587308e-2	8.587284e-2	8.587289e-2	8.587284e-2
2	960	5.262487e-2	5.262505e-2	5.262490e-2	5.262526e-2
3	3840	3.251092e-2	3.251115e-2	3.251097e-2	3.251136e-2
4	15360	2.021560e-2	2.021575e-2	2.021563e-2	2.021590e-2
5	61440	1.262635e-2	1.262645e-2	1.262637e-2	1.262655e-2
6	245760	7.909156e-3	7.909221e-3	7.909171e-3	7.909282e-3

In Examples 1, 3, and 4 the difference in the accuracy for all four u_h^i, $1 \leq i \leq 4$, is smaller than 0.1% on the finest refinement levels. Only in Example 2 can a significant difference be observed. In that case, the errors in the L^2-norm differ by more than 100% between u_h^1 and u_h^3. Additional test examples with singularities show that no clear pattern can be observed.

The following two figures illustrate the discretization errors given in Tables 1.1–1.6 for $u - u_h^1$ (standard) and $u - u_h^3$ (dual). In Fig. 1.32, the errors in the energy norm are visualized whereas in Fig. 1.33 the errors in the L^2-norm are shown. In each figure a straight dashed line is drawn below the obtained curves to indicate the asymptotic behavior of the discretization errors.

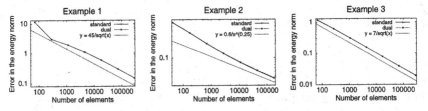

Fig. 1.32. Discretization errors in the energy norm versus number of elements

In Examples 1 and 3, we can observe the predicted order h for the energy norm and the order h^2 for the L^2-norm almost from the beginning. In these two examples only one plotted curve for the standard Lagrange multiplier space M_h^1 and the dual space M_h^3 can be seen, since the numerical results are too close. In Example 2, where we have no full H^2-regularity, the asymptotic behavior starts late. We observe an $\mathcal{O}(h^{1/2})$ behavior for the discretization errors in the energy norm for both mortar methods. During the first refinement steps the error decreases more rapidly than later. For the L^2-norm the asymptotic rate is $\mathcal{O}(h^{3/2})$. Moreover, it seems to be the case that the Lagrange multiplier space M_h^3 performs asymptotically better than the standard one. However, this has not been observed for some other examples without full regularity. The discretization error in the L^2-norm is

more sensitive to the choice of the Lagrange multiplier space than the error
in the energy norm.

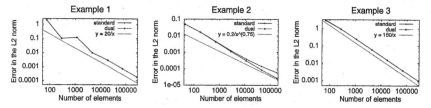

Fig. 1.33. Discretization errors in the L^2-norm versus number of elements

We remark that the theory does not make any statements about the
constants in the a priori estimates. However, the numerical results show that
the discretization errors for the different Lagrange multipliers are almost
the same. Thus from the point of accuracy, there is no preferable Lagrange
multiplier space.

1.5.2 A Non-optimal Mortar Method

To get a better understanding of the situation at the interface and the role of
the Lagrange multiplier, we now consider a non-optimal discretization. Here,
we understand non-optimal in the sense that no order h a priori estimate
can be obtained. Such a discretization scheme can easily be constructed if
we use different Lagrange multiplier spaces on the two sides of the interface.
We like to use a biorthogonal basis on the non-mortar side, and nonnegative
Lagrange multiplier basis functions on the mortar side. We start with the
spaces M_h^2 and M_h^3 and define the constrained space $\widehat{V}_h$ by

$$\widehat{V}_h := \left\{ v \in X_h \mid \int_{\gamma_m} (v_{|\Omega_{n(m)}} \psi_i^3 - v_{|\Omega_{\overline{n}(m)}} \psi_i^2)\, d\sigma = 0,\ 1 \le m \le M,\ i \le N_m \right\} ,$$

where ψ_i^2 and ψ_i^3 are defined as in Subsect. 1.2.4. We recall that ψ_i^2 and ψ_i^3 are
both associated with the same vertex on γ_m. Using the same tools as discussed
in detail in Sect. 1.1 and 1.2, it can be verified that the consistency error is
of order h but not the best approximation error. For the best approximation
error, we find only

$$\inf_{v_h \in \widehat{V}_h} \|u - v_h\|_1^2 \le C \left(\sum_{k=1}^{K} h_k^2 |u|_{2;\Omega_k}^2 + \sum_{m=1}^{M} h_m |u|_{1;\gamma_m}^2 \right) . \tag{1.65}$$

We do not prove this result because it is of no theoretical interest. However,
it can easily be seen that no optimal result can be obtained.

Let us assume that the triangulation is not uniform in the sense that at least one vertex x_i at an interface γ_m, $1 \leq m \leq M$, not adjacent to the endpoints of γ_m, is not at the same time at the midpoint m_i of the support of ψ_i^2; see Fig. 1.34.

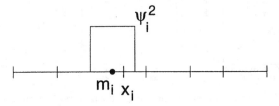

Fig. 1.34. Vertex x_i and midpoint m_i do not coincide

Then, for $v \in P_1(\Omega)$, we obtain

$$v(x_i) = \frac{1}{|\text{supp}\psi_i^2|} \int\limits_{\gamma_m} v\psi_i^3 \, d\sigma, \quad v(m_i) = \frac{1}{|\text{supp}\psi_i^2|} \int\limits_{\gamma_m} v\psi_i^2 \, d\sigma \ .$$

If v is not constant along γ_m, then $v(x_i) \neq v(m_i)$ since by assumption $x_i \neq m_i$. From the definition of $\widehat{V}_h$, we find that $P_1(\Omega) \not\subset \widehat{V}_h$. Here, we have assumed that no Dirichlet boundary condition is imposed on the product space X_h. In the case of such a triangulation, $P_1(\Omega)$ is not a subspace of $\widehat{V}_h$, and no a priori bound in terms of $h|u|_2$ holds.

Based on this observation one might get the idea of replacing ψ_i^2 in the definition of $\widehat{V}_h$ by $\widehat{\psi}_i$, where $\widehat{\psi}_i$ is equal one in a circle with midpoint x_i and a radius depending on the local meshsize, and zero outside. From the point of view of the approximation quality this creates an issue. In fact, such a modification saves the best approximation property but the optimal order of the consistency error is then lost unless we again have a very special triangulation.

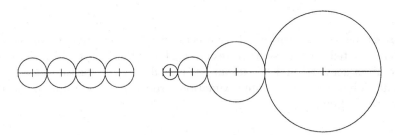

Fig. 1.35. Special triangulations on the interface

The approximation property requires that x_i is the midpoint of the support of $\widehat{\psi}_i$, and the analysis for the consistency error requires that the sum over the basis functions on the interfaces equals one. Figure 1.35 shows two example of triangulations for which consistency and approximation property can be satisfied at the same time. The circles in Fig. 1.35 indicate that the vertices are the midpoints of the supports of the corresponding basis functions. We remark that only the midpoints of the circles are vertices. At the left, two adjacent circles intersect at the midpoint of the edges. This is not the case on the right, there the ratio of the length of two adjacent edges, not sharing an endpoint of γ_m, is the same but not equal to one. Only if all triangulations associated with the non-mortar sides have a structure as indicated in Fig. 1.35, has the constrained space $\widehat{V}_h$ with ψ_i^2 replaced by $\widehat{\psi}_i$ an order h approximation property. If this condition is violated, we cannot, according to the theory, expect an order h a priori estimate of the discretization error in the energy norm.

The duality of the requirements for the approximation property and the consistency error indicates that the use of different spaces as test and trial space could be an issue for this problem: Find $\hat{u}_h \in \widehat{V}_h$ such that

$$a(\hat{u}_h, v_h) = (f, v_h)_0, \quad v_h \in \widetilde{V}_h \ .$$

A similar situation is studied in [CLM97]. In that paper, one Lagrange multiplier space is used but the exact integral over the interfaces is replaced by numerical quadrature. One quadrature rule is based on the mesh on the non-mortar side and the other on the mesh on the mortar side. Then, again the approximation property and the consistency error are optimal only for very special triangulations. The idea is now to use a variational formulation where test and trial spaces are different. No a priori estimates for the discretization error are proved. In particular, the well-posedness of the variational problem is an open question. However, numerical results indicate that it might be an order h method; see [CLM97].

The following two tables show the discretization errors for this non-optimal approach for the examples already discussed. The initial triangulations do not have the required form in any of the Examples 1–4. However, if we compare the numerical results with those obtained in Subsect. 1.5.1, we get almost the same accuracy.

Table 1.9. Discretization errors in the L^2-norm for Examples 1–4

level	Example 1	Example 2	Example 3	Example 4	$u = 3x + 2y$
0	2.021200e-0	4.855250e-2	2.981412e-0	6.530118e-3	9.961053e-3
1	1.019843e-1	1.650063 e-2	9.358003e-1	3.296302e-3	7.211486e-3
2	1.170394e-1	4.380014e-3	2.431747e-1	1.158950e-3	1.990688e-3
3	9.532570e-3	1.166031e-3	6.104268e-2	4.268715e-4	5.424928e-4
4	2.801035e-3	3.234870e-4	1.525571e-2	1.628104e-4	1.454180e-4
5	7.115364e-4	9.563012e-5	3.812352e-3	6.351176e-5	3.824237e-5
6	1.786010e-4	3.041376e-5	9.529113e-4	2.504495e-5	9.944375e-6

Table 1.10. Discretization errors in the energy norm for Examples 1–4

level	Example 1	Example 2	Example 3	Example 4	$u = 3x + 2y$
0	1.147916e+1	6.045399e-1	1.199261e-0	1.410401e-1	1.806513e-1
1	3.029716e-0	3.613288e-1	6.187424e-1	8.591741e-2	2.085708e-1
2	1.953580e-0	2.076923e-1	3.094936e-1	5.264523e-2	9.902825e-2
3	1.117625e-0	1.254030e-1	1.546515e-1	3.252204e-2	5.020210e-2
4	5.928700e-1	7.980953e-2	7.731113e-2	2.022203e-2	2.521881e-2
5	2.979264e-1	5.300016e-2	3.865380e-2	1.263017e-2	1.256562e-2
6	1.491143e-1	3.619913e-2	1.932670e-2	7.911474e-3	6.231056e-3

In Examples 2 and 3, the initial triangulations on the non-mortar side are almost of the required form. The only exceptions are the vertices adjacent to the endpoints of the non-mortar sides. Only in Examples 1 and 4, are the triangulations not as close to the required form. In Example 1, we get minimal better results, and in Example 4, we obtain minimal worse results than before. We again observe that the L^2-error is more sensitive to the choice of the constraints than the energy error. These numerical results would indicate that the discretization error in the energy norm is of order h and in the L^2-norm of order h^2. The effect of the non-optimality of the method is shown in Column 6. We use the Laplace operator, select $u = 3x + 2y$ as exact solution, and choose the same decomposition into subdomains as in Example 1 but the triangulations on the non-mortar sides are more irregular. An optimal method would yield zero as discretization error on each level which is obviously not the case for this setting.

1.5.3 Influence of the Choice of the Mortar Side

The following numerical results show that the choice of the non-mortar sides and the appropriate refinement at the interfaces are quite crucial in the case of discontinuous coefficients. The energy norm as well as the L^2-norm are considered for two different situations. Uniform and adaptive refinement techniques are used to illustrate the influence of the choice of the mortar side. In the case of an adaptive scheme, we start with a conforming triangulation, whereas in the case of uniform refinement, we use a highly non-matching initial triangulation.

We consider $-\text{div}(a\nabla u) = f$ on Ω. The coefficient a is constant on the different subdomains and has a jump across the interfaces. We consider only examples where the right hand side f does not reflect the jump in the coefficient a. For examples where f reflects the jump, there would be a principal difference in the results. In Situation I, the non-mortar side is defined where the coefficient a is smaller whereas in Situation II, the non-mortar side is associated with the larger coefficient. We start with an example of highly non-matching triangulation and strongly discontinuous coefficients. Figure 1.36 shows the decomposition into subdomains, the non-matching triangulations and the isolines of the numerical solutions for Situations I and II. The coefficient a has the same value on the inner and outer subdomain and is smaller in the middle one. Although in both situations, the same triangulation is used, two completely different results are obtained. A good approximation can be found only in Situation I when M_h is of higher dimension. The constrained space V_h is of higher dimension in Situation II, but then, the approximation is incorrect; see [HIK+98, BDL99].

 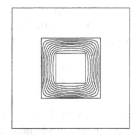 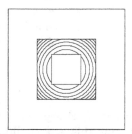

Fig. 1.36. Triangulation (left), solution for Sit. I (middle) and Sit. II (right)

In the rest of this subsection, we focus on the influence of the choice of the non-mortar side. We study this influence in Examples 5 and 6 for adaptive and uniform refinement strategies. To start, we briefly discuss some aspects of an adaptive refinement algorithm. For standard conforming discretizations, different techniques for the construction of efficient and reliable local a posterior error estimators are well known. Residual based error estimators or hierarchical basis estimators based on a higher order discretization scheme are very often used. We refer to [Ver96] for an excellent overview and introduction to the basic concepts. Both residual and hierarchical error estimators have been adapted to the mortar settings [BH99, PS96, Woh99a, Woh99c]. The main problem is to take care of the non-nestedness of the nonconforming finite element spaces. It turns out; see Lemma 1.7 in Sect. 1.2, that an appropriate measure for the nonconformity is a weighted L^2-norm of the jump. In particular, it can be shown that an upper bound for this weighted L^2-norm is given by the discretization error in the energy norm. For each element, with an edge on a non-mortar side, the jump term

$$\frac{a_e}{h_e}\|[u_h]\|_{0;e}^2$$

is part of the elementwise contribution of the error estimator. Here, a_e stands for the coefficient on the non-mortar side.

In the case of standard conforming triangulations, certain elements are marked by the error estimator and refined, and the refinement rules create a conforming triangulation in each adaptive step. In contrast to this, no additional rules control the adaptive refinement at the interfaces in the mortar settings. The only information transfer between the subdomains is obtained by the local contributions of the error estimators. A local error estimator which does not reflect the jump can therefore not guarantee an appropriate refinement at the interface. The error in the Dirichlet boundary conditions at the interfaces is measured by the jump term of the finite element solution, while the difference between the discrete flux and the discrete Lagrange multiplier $a \nabla u_h \cdot \mathbf{n} - \lambda_h$ controls the error for Neumann boundary conditions. A detailed discussion of a posteriori techniques for mortar finite elements can be found in [BH99, PS96, Woh99a, Woh99c].

Here, we choose a mean value strategy to control the adaptive refinement process. An element $T \in \mathcal{T}_l$ on level l is marked for refinement in the next step if the local contribution of the error estimator η_T satisfies

$$\eta_T^2 \geq \sigma \frac{1}{N_l} \sum_{T' \in \mathcal{T}_l} \eta_{T'}^2 \, ,$$

where N_l is the number of elements in $\mathcal{T}_l$, and we set $\sigma = 0.95$. The idea behind this is to equilibrate the error per element and to obtain a prescribed accuracy at a minimal cost.

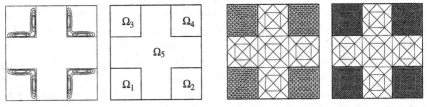

Fig. 1.37. Isolines of the solution (left), decomposition into 5 subdomains (left middle), level difference 2 (right middle) and level difference 3 (right) of the initial triangulation, (Example 5)

Example 5 illustrates the influence of the choice of the Lagrange multiplier. We consider the diffusion equation $-\operatorname{div} a \nabla u = f$, on $(0,1)^2$, where the coefficient a is discontinuous. This example is discussed in detail in [Woh99a], where hierarchical error estimators for mortar methods are studied. The unit square Ω is decomposed into five subdomains as shown in Fig. 1.37. The coefficients in the subdomains Ω_i are given by $a_5 = 5000$, $a_i = 1$, $i \in \{1, 2, 3, 4\}$. The right hand side f and the Dirichlet boundary conditions are chosen to match an exact solution, $u(x_1, x_2) = 1/a \, \sin(3\pi x_1) \sin(3\pi x_2) \sum_{i,j=1}^{2} \exp(-800(x_j - i/3)^2)$. This solution is continuous and $[a_i \nabla u \cdot \mathbf{n}]_J$ is equal to zero on the interfaces. The isolines of the solution are shown in Fig. 1.37. We now consider the two different possible choices of the mortar sides separately. In Situation I, the discrete Lagrange multiplier space is associated with the triangulations given on Ω_1, Ω_2, Ω_3 and Ω_4, whereas in Situation II, the triangulation for the Lagrange multiplier space is inherited from the one on Ω_5.

Figure 1.38 shows the influence of the choice of the Lagrange multiplier on the adaptive refinement process for Example 5; the adaptively refined triangulation on Level 3 and Level 4 are given for both situations. In Situation I, we observe a sharp interface between the different subdomains. However this is not the case for Situation II, where we obtain a triangulation which tends to be more conforming at the interfaces. Furthermore, more nodes are generated on the side where a is larger. The additional refinement is a consequence of the choice of the non-mortar side.

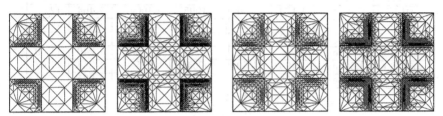

Fig. 1.38. Situation I (left) and Situation II (right), (Example 5)

We recall that the dimension of the Lagrange multiplier space is given by the numbers of edges on the non-mortar side minus one. A sharp interface in Situation II would lead to poor approximation properties for the Neumann boundary condition. Thus, we have to adapt the triangulation along the interface on the side where a is larger, and increase the dimension of the Lagrange multiplier space; see the right part of Fig. 1.38. However, the number of these additional elements next to the interface can be neglected asymptotically because they are associated with a 1D interface problem.

In Situation II, the jump of the discrete solution plays an important role in the definition of the error estimator, since it controls the nonconformity of the mortar finite element solution. Without this term the error estimator would fail. In Situation I, it could be neglected and no significant difference can be observed in the adaptive triangulations. We point out that the adaptive refinement on both sides of the interface in Situation II is not enforced by any refinement rules but only by the local contributions of the error estimator.

Comparing the true error, in Table 1.11, we find that the performance is asymptotically the same in Situations I and II, i.e., the number of elements, to obtain a given accuracy, is asymptotically almost the same. We introduce

$$\chi := \sqrt{N_{\mathcal{T}}} \, \|u - u_h\|$$

as measure of the performance. In fact, χ gives a rough idea of how many elements $N_{\mathcal{T}}$ are required to obtain a given accuracy; the smaller χ, the fewer elements are required. The ratio in Situation II between the number of additional elements in the neighborhood of the interfaces and the total number of elements tends to zero in the adaptive refinement process.

Table 1.11. Effectivity index ζ and performance χ, (Example 5)

Situation I				Situation II			
# elem.	true err.	ζ	χ	# elem.	true err.	ζ	χ
144	1.771e-0	1.674	21.25	144	1.771e-0	1.996	21.25
232	1.285e-0	1.301	19.57	312	1.285e-0	2.572	22.71
456	6.302e-1	1.244	13.46	536	7.789e-1	1.388	18.03
1016	3.794e-1	1.192	12.09	1112	4.563e-1	2.826	15.22
2632	2.309e-1	1.155	11.85	1600	3.715e-1	1.258	14.86
6088	1.605e-1	1.140	12.52	3688	2.155e-1	1.913	13.08
12880	1.024e-1	1.159	11.62	4996	1.805e-1	1.223	12.76
31012	6.789e-2	1.170	11.96	11905	1.100e-1	1.404	12.00
57344	4.600e-2	1.226	11.02	19700	8.392e-2	1.241	11.78
138184	3.156e-2	1.179	11.73	44962	5.430e-2	1.271	11.51
240360	2.228e-2	1.244	10.92	88471	3.896e-2	1.240	11.59
568780	1.525e-2	1.199	11.50	190545	2.535e-2	1.248	11.07
972488	1.103e-2	1.252	10.88	428246	1.774e-2	1.235	11.61

In our experiment, we start with a global conforming triangulation. Each subdomain is decomposed into 16 elements. In the first refinement steps, we observe that χ is considerably larger in Situation II. However, this difference vanishes asymptotically. From Level 8 on, χ can be seen to oscillate. Asymptotically it appears as if χ is smaller for Situation I than Situation II, when the level is even, while the opposite is true on an odd level. This is also reflected in the effectivity index ζ which is defined by

$$\zeta := \frac{\text{estimated error}}{\text{true error}} \, ,$$

and which is a measure of the quality of the error estimator. For a good estimator, it is required that ζ tends asymptotically to a value close to one.

In a second test setting for the same example, we start with a highly non matching triangulation at the interface and use uniform refinement techniques. We work with level differences of two and three between the triangulations on the subdomains; see Fig. 1.37 for the initial triangulations. The choice of the initial triangulation is motivated by the following observation: Within the adaptive refinement process, we find in Situation I a level difference of approximately 3 at the interface and thus, roughly $h_1/h_2 \approx \sqrt[4]{a_1/a_2}$. This reflects the fact that during the refinement process the error in the energy norm per element is equilibrated.

In Tables 1.12 and 1.13, the discretization errors for the two different situations are given. Table 1.12 shows the discretization errors in the energy norm as well as in the L^2-norm for Situation I and Situation II, if the initial triangulation has a level difference of three at the interface; see the right picture in Fig. 1.37. In Situation II, the results are significantly worse. Columns 6 and 7 give the ratio between the errors. For the L^2-norm, the ratio between the errors in the two situations is almost ten, even after four refinement steps.

The errors in Situation II are extremely bad compared to those of Situation I. For the energy norm this ratio is not as extreme and improves considerable with an increasing number of nodes. Asymptotically, it seems to tend to one.

Table 1.12. Discretization errors in the case of a level difference 3, (Example 5)

# elem.	Situation I		Situation II		Ratio I/II	
	En-error	L^2-error	En-error	L^2-error	En	L^2
4176	2.8184e-1	1.2120e-3	1.0216e-0	3.9234e-2	0.276	0.031
16704	1.4485e-1	3.4809e-4	4.4993e-1	8.7215e-3	0.322	0.041
66816	7.0739e-2	8.2695e-5	1.5808e-1	1.5457e-3	0.477	0.054
267264	3.5318e-2	2.0606e-5	6.2716e-2	3.0342e-4	0.563	0.068
1069056	1.7654e-2	5.1475e-6	2.2741e-2	4.2456e-5	0.776	0.121

Table 1.13. Discretization errors in the case of a level difference 2, (Example 5)

# elem.	Situation I		Situation II		Ratio I/II	
	En-error	L^2-error	En-error	L^2-error	En	L^2
1104	5.5021e-1	4.5473e-3	1.1269e-0	3.9831e-2	0.488	0.114
4416	2.8155e-1	1.2254e-3	4.9749e-1	8.3860e-3	0.566	0.146
17664	1.4002e-1	3.0974e-4	1.9554e-1	1.5039e-3	0.716	0.206
70656	6.9964e-2	7.7758e-5	8.6293e-2	2.9904e-4	0.811	0.260
282624	3.4976e-2	1.9458e-5	3.7670e-2	4.4813e-5	0.929	0.434
1130496	1.7487e-2	4.8657e-6	1.7848e-2	7.0931e-6	0.980	0.686

The same type of results is given in Table 1.13, but now, the initial triangulation has only a level difference of two at the interface instead of three. The difference between Situations I and II is not as extreme as in Table 1.12. Starting with a ratio for the energy norm of approximately 0.5 on the initial triangulation, the ratio tends to one with an increasing number of refinement levels; on the finest level, where we have more than a million elements, the ratio is approximately 0.98. We observe the same type of behavior for the L^2-norm. However, even more nodes are needed before the asymptotic ratio of one is reached; the L^2-norm depends more sensitively on the choice of the Lagrange multiplier than the energy norm.

Finally, Fig. 1.39 displays the numbers of Table 1.12 and Table 1.13. For Situation I, the correct order of the discretization scheme can be observed from the beginning. All curves parallel to $y = 15/\sqrt{x}$ and $y = 3/x$ have the correct order in the energy norm and the L^2-norm, respectively. It can be seen that the adaptive refinement on both sides of the interface for Situation II, shown in Fig. 1.38, is really necessary. A sharp interface between the subdomains makes sense only for Situation I. In Situation I, there is almost no difference in the accuracy obtained if a level difference of 2 or 3 is used. However in Situation II, a level difference of 3 yields much worse results. For the energy norm, it appears that asymptotically we get the same performance for the different choices. However, the asymptotic behavior starts very late and depends sensitively on the level difference.

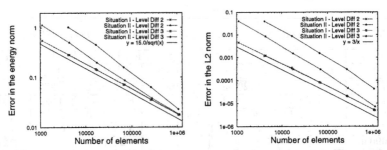

Fig. 1.39. Error in the energy (left) and in the L^2-norm (right), (Example 5)

For the L^2-norm the observed phenomena are even more significant, and the obtained accuracy for a highly non-matching triangulation depends strongly on the choice of the non-mortar side. In the case of a level difference 3, the asymptotic range will never be reached in practical relevant computations for the L^2-norm. Thus, working with highly non-matching triangulations and uniform refinement techniques requires the proper choice of the non-mortar side. The error estimator can deal with both situations. In Situation I, a sharp interface will be generated, whereas in Situation II adaptive refinement will be observed on both sides of the interface.

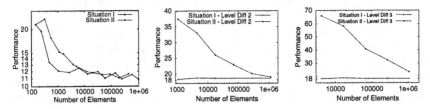

Fig. 1.40. Performance adaptive (left) and uniform refinement, level difference 2 (middle) and level difference 3 (right), (Example 5)

Figure 1.40 shows the performance of the different refinement strategies and situations. The adaptive strategy has a considerably better performance than the uniform strategy. For both situations the performance tends asymptotically to values between 11 and 12 in the adaptive case. In Situation I, the asymptotic range is reached earlier than in Situation II, and the performance is slightly better. The asymptotic behavior starts after a few refinement steps for both situations. In the middle and right part of Fig. 1.40, the performance for the uniform case is displayed. Here, we observe a big difference between Situation I and Situation II. We consider the cases of two (Case 1) and three

(Case 2) level differences in the initial triangulations; see Fig. 1.37. If the mortar side is chosen appropriately, i.e., we are in Situation I, we obtain an almost constant performance from the beginning on. In Situation I, we obtain a performance of ≈ 18.25 in Case 2 and a performance of ≈ 18.59 in Case 1. The performance of Situation II, seems to tend asymptotically to the performance of Situation I. But the asymptotic starts extremely late. In Case 1, the asymptotic is reached after five refinement steps. In Case 2, the asymptotic is not reached even for more than a million elements. The bad approximation property of the Lagrange multiplier in Situation II results in the poor performance. However, the approximation property of the Lagrange multiplier space is of order $h_{\mathrm{non}}^{3/2}$, where h_{non} is the meshsize on the non-mortar side. In Situation II, h_{non} is considerably larger than the meshsize on the mortar side. The fact that the approximation order of the Lagrange multiplier space is higher than the approximation order of the finite element space yields that the influence of the choice of the mortar side vanishes asymptotically. But the asymptotic starts so late, that it will be not reached for many numerical applications. Using uniform refinement strategies, it is extremely important for the performance of the method to make the correct choice of the mortar side.

In our last example in this subsection, we consider a different coefficient in each subdomain. It has been originally studied in [Woh99b]. As in Example 5, we find an essential difference between the two choices of mortar sides. In the case of uniform refinement, the obtained accuracy depends highly on this choice and in the case of adaptive refinement, the refinement at the interface reflects this choice.

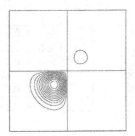 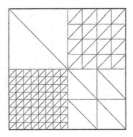

Fig. 1.41. Isolines of the solution (left) and non-matching initial triangulation (right), (Example 6)

The unit square is decomposed into four subdomains $\Omega_{ij} := (i/2, (i+1)/2) \times (j/2, (j+1)/2)$, $i, j \in \{0, 1\}$ and $a|_{\Omega_{ij}} := a_{ij}$ where $a_{00} := 1$, $a_{10} := 250$, $a_{01} := 5000$ and $a_{11} := 10$. The data are chosen to match the given solution $u(x, y) = (x - 0.5)(y - 0.5) \exp(-40(x - 0.5)^2) \exp(-40(y - 0.5)^2)/a$. The isolines of the solution are shown in Fig. 1.41. Figure 1.42 shows the adaptive refinement process for both situations on Level 4 and Level 5. We use a conforming initial triangulation with 8 elements. As in Example 5, we obtain a sharp interface between the subdomains in Situation I whereas in Situation II, adaptive refinement is observed on both sides of the interface.

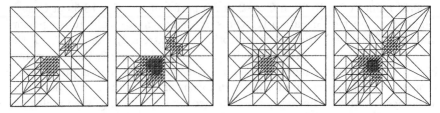

Fig. 1.42. Adaptive refinement for Situations I (left) and II (right), (Example 6)

Finally, we compare Situations I and II for a given triangulation with a sharp interface between highly refined and unrefined subdomains. On each subdomain a uniform triangulation is used; see Fig. 1.41. The meshsizes in the subdomains reflect the ratio of the coefficient on the subdomains. Although the constrained finite element space has more degrees of freedom in Situation II, the discretization errors in the energy norm as well as in the L^2-norm are worse compared with Situation I. Due to the bad approximation property of the Lagrange multiplier in Situation II, the consistency error of the discretization appears to play a significant role.

Table 1.14. Error in the energy and L^2-norm for Situations I and II in the case of uniform refinement, (Example 6)

# elem.	Situation I		Situation II		Ratio I/II	
	En-error	L^2-error	En-error	L^2-Error	En	L^2
170	3.6023e-3	4.4479e-5	6.1857e-3	2.4307e-4	0.582	0.183
680	1.8419e-3	1.1449e-5	3.2997e-3	7.0047e-5	0.558	0.163
2720	9.2500e-4	2.8676e-6	2.0115e-3	3.4793e-5	0.460	0.082
10880	4.6427e-4	7.1992e-7	7.4883e-4	4.5648e-6	0.620	0.158
43520	2.3259e-4	1.8059e-7	3.1681e-4	7.2615e-7	0.734	0.249
174080	1.1637e-4	4.5265e-8	1.4665e-4	1.4541e-7	0.793	0.311

In Table 1.14, the errors in the L^2-norm as well as in the energy norm are given. Columns 6 and 7 show the ratio of the discretization errors for the two different situations. As in Example 5, the error in the L^2-norm depends in a more sensitive way on the choice of the mortar side than the error in the energy norm. Asymptotically the ratio seems to tend to one. However, with more than 100000 elements the ratio in the L^2-norm is still only 0.311.

With a sharp interface and uniform refinement, satisfying results can only be obtained for Situation I. Applying adaptive refinement strategies, both situations can successfully be handled, and two different types of refined triangulations will be generated according to the choice of the mortar sides.

1.5.4 Influence of the Jump of the Coefficients

Our last example in this section, reflects the influence of the jump in the coefficient on the adaptive refinement process. Adaptive refined triangulations are compared for different jumps in a. We consider $-\text{div}(a\nabla u) = f$ on the unit square. The right hand side f and the Dirichlet boundary conditions are chosen for an exact solution $u = \exp(-1500 * (r^2 - 0.2)^2) - \exp(-3000 * (r^2 - 0.075)^2)$ for $a = 1$, where $r^2 := (x - 0.5)^2 + (y - 0.5)^2$. Here, Ω is decomposed into nine subdomains $\Omega_{ij} := (i/3, (i+1)/3) \times (j/3, (j+1)/3), 0 \leq i, j \leq 2$. The coefficient a is now chosen piecewise constant using a red and black ordering of the subdomains: $a := a_1$ on Ω_{ij} if $i+j$ even and $a := a_2$ on Ω_{ij} if $i+j$ odd.

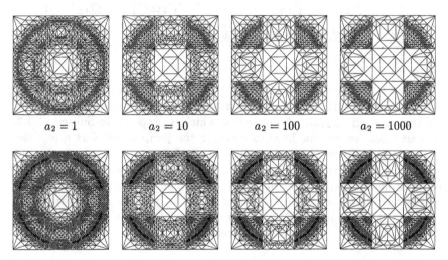

$a_2 = 1$ $\qquad$ $a_2 = 10$ $\qquad$ $a_2 = 100$ $\qquad$ $a_2 = 1000$

Fig. 1.43. Adaptive refinement on Level 4 and Level 5, (Example 7)

Here, the influence of the jump in the coefficients on the generated triangulations is of interest. Crouzeix–Raviart elements on Ω_{CR} and conforming P_1-elements on Ω_{P_1}, $\overline{\Omega} = \overline{\Omega}_{P_1} \cup \overline{\Omega}_{CR}$, are coupled by means of piecewise constant Lagrange multipliers; as in Subsect. 1.4.2. The Crouzeix–Raviart elements are used in the subdomains where $a = a_1$. Figure 1.43 shows the adaptively generated triangulations on Level 4 and Level 5 for the choice of $a_1 := 1$ and $a_2 \in \{1, 10, 100, 1000\}$.

With increasing a_2 the obtained triangulations tend to be more and more nonconforming at the interfaces between the subdomains. We recall that no matching condition is imposed for the triangulations of the different subdomains. Considering the situation at the interfaces in more detail, we observe that the jump between the meshsizes depends on the jump of the coefficient a. In the case of $a_2 = 1$, we obtain an almost matching triangulation at the interfaces, whereas in the case of $a_2 = 1000$ a highly non uniform triangulation is generated. Roughly speaking, we find that the ratio of the local meshsizes h_1/h_2 is approximately $\sqrt[4]{a_1/a_2}$. This is related to the mean value refinement strategy being used.

Table 1.15. Effectivity index on Ω_{P_1}, Ω_{CR} and Ω $(a_2 = 1)$, (Example 7)

level	# elem.	ζ on Ω_{P_1}	ζ on Ω_{CR}	ζ on Ω	err. Ω_{P_1}/Ω_{CR}
1	144	1.2121	2.3357	1.8060	1.106
2	528	1.1487	1.4936	1.3543	0.965
3	1408	1.0323	1.3847	1.1981	1.042
4	1864	1.1370	1.4660	1.3065	1.033
5	4048	1.2573	1.5403	1.4070	0.992
6	9032	1.2756	1.5415	1.4189	0.996
7	21112	1.2763	1.5784	1.4340	1.009
8	43944	1.3049	1.6090	1.4609	1.026
9	100600	1.3105	1.6076	1.4704	0.974

Table 1.15 shows the effectivity index related to Ω as well as to the two different discretizations. We use the simplified residual type error estimator for the Crouzeix–Raviart discretization which is proposed in Subsect. 1.4.3. In particular, we do not have to compute the jump of the flux across the edges. It is sufficient to evaluate the nonconformity of the finite element solution. For the conforming P_1-elements, we apply a scaled residual based error estimator; see, e.g., [Ver96]. In both cases, we observe an overestimation of the error. The last column reflects the ratio of the error in the energy norm on the subdomains Ω_{P_1} and Ω_{CR}. Asymptotically, it is close to one alternating between values less and greater than one.

Fig. 1.44. Adaptive refinement on Level 6 for $a_2 = 0.1$ (left) and $a_2 = 0.01$ (right)

Finally, Fig. 1.44 shows the triangulations on Level 6 for $a_2 \in \{0.1, 0.001\}$. Now, the non-mortar sides are defined on Ω_{ij} with $i + j$ odd. As before, we observe a sharp interface between the different subdomains.

2. Iterative Solvers Based on Domain Decomposition

This chapter concerns iterative solution techniques for linear systems of equations arising from the discretization of elliptic boundary value problems. Very often huge systems are obtained, with condition numbers which depend on the meshsize h of the triangulation, which typically grow in proportion to h^{-2}. Then, classical iteration schemes like Jacobi-, Gauß–Seidel or SOR-type methods result in very slow convergence rates. Figure 2.1 shows the convergence rates and the number of iteration steps versus the number of unknowns, for a simple model problem in 2D. In the left, the convergence rates are given, and in the right the number of iteration steps to obtain an error reduction of 10^{-6} are shown. For the Jacobi and the Gauß–Seidel method, the asymptotic convergence rates are $1 - \mathcal{O}(h^2)$. The optimal SOR-method is asymptotically better and tends with $\mathcal{O}(h)$ to one. However, the optimal damping parameter is, in general, unknown. The number of required iteration steps reflects the order of the method. For the Gauß–Seidel and the Jacobi method, the number of required iteration steps grows quadratically with one over the meshsize. In case of the optimal SOR-method, the increase is linear. Moreover, the numerical results show that the Jacobi method requires two times the number of Gauß–Seidel iteration steps. In case of the optimal SOR-method

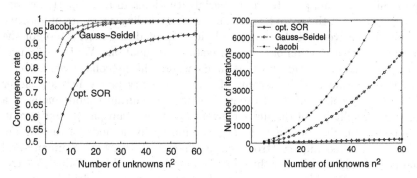

Fig. 2.1. Convergence rates and number of iterations

Nonlinear iteration schemes as the conjugate gradient method give better results with a convergence rate depending on the square root of the condition number of the preconditioned system. Figure 2.2 illustrates the quality of a preconditioned conjugate gradient methods in 2D. A hierarchical basis method and the BPX-preconditioner are tested; we refer to [BPX90a, BPX90b, Yse86, Yse93] and Subsect. 2.1.1. If a conjugate gradient method is applied without preconditioner, the number of iteration steps to obtain a given accuracy is inversely proportional to the meshsize. In the case that a preconditioned version is used, the required number of iteration steps can be much smaller, and is in the best case independent of the meshsize. For the hierarchical basis method, we observe a logarithmic growths in 2D and the BPX-preconditioner results in an optimal method.

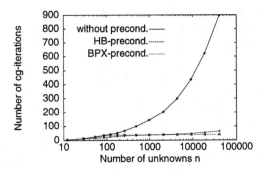

Fig. 2.2. Convergence rates of preconditioned cg-method

Here, we focus on the construction of efficient iterative solvers for non standard discretization schemes. In particular, vector fields discretization techniques such as Raviart–Thomas finite elements, Nédélec finite elements, and mortar methods are considered. Well established techniques for Lagrangian P_1-elements such as iterative substructuring or multigrid methods are modified and adapted to these special situations. In a first part, we consider the abstract theory of Schwarz methods including examples of additive and multiplicative variants. These techniques provide a powerful tool for the efficient iterative solution of the huge systems of equations. We focus on preconditioned conjugate gradient methods where the preconditioner is built from the solution of subproblems of less complexity and which are either related to a decomposition of the geometrical domain into subdomains or a hierarchical splitting of the finite element space into subspaces. Our second main concern is the construction of special multigrid methods for the domain decomposition techniques introduced in Chap. 1. A Dirichlet–Neumann and two different multigrid algorithms for the mortar method are studied. We can interpret the Dirichlet–Neumann method as a block Gauß–Seidel preconditioner for the unsymmetric mortar formulation. The first of the proposed multigrid method is based on the new positive definite mortar formulation on the unconstrained product space whereas the second one works with the saddle point formulation.

2.1 Abstract Schwarz Theory

In this section, we give only a brief overview of the general framework of Schwarz methods. Many classes of preconditioners for large linear systems of equations arising from the discretization of partial differential equations have been analyzed within this framework; see, e.g., [BS94, Le 94, QV99, SBG96, Wid99] and the references therein. Applications are particularly well developed for conforming finite element approximations of elliptic problems. For details, we refer to [BPWX91a, CM94, DW95, Osw94, SBG96, Xu92, Yse93]. Technical tools for establishing upper bounds for the condition number are given and briefly discussed in this section. We outline examples such as a two-level overlapping method [Bre00, DW94], an iterative substructuring method [BPS86b, BPS89, XZ98], and a multilevel method [Ban96, Yse86, Yse93, Xu92] for standard conforming piecewise linear finite elements.

The possibly earliest domain decomposition method appears to be that of Hermann A. Schwarz [Sch90]. He introduced an alternating method to prove the existence of harmonic extensions on domains with nonsmooth boundaries, more than one hundred years ago. Figure 2.3 shows the decomposition of such a domain into a circle and a rectangle; this type of decomposition is used in the original work of Schwarz [Sch90] to illustrate his idea. A sequence of harmonic functions is constructed in each subdomain and is shown to converge; the convergence rate depends on the overlap. The original proof was based on the maximum principle. We refer to [QV99] for a convergence proof of the alternating Schwarz method based on the variational framework; this approach is introduced in [Lio88].

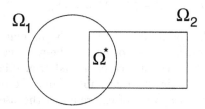

Fig. 2.3. Decomposition used by Schwarz in his original work

Here, we briefly review the idea of Schwarz in the context of an elliptic operator L. Let Ω_1 be a circle and Ω_2 be a rectangle as shown in Fig. 2.3. The intersection between Ω_1 and Ω_2 is denoted by Ω^*. To prove the existence of a function satisfying

$$Lu = f, \quad \text{in } \Omega, \ \overline{\Omega} := \overline{\Omega}_1 \cup \overline{\Omega}_2 \ ,$$
$$u = g, \quad \text{on } \partial\Omega \ ,$$

we proceed as follows: Assuming the existence of solutions of the elliptic equation on the subdomains Ω_1 and Ω_2 for given suitable boundary conditions, we obtain a sequence alternatingly updated on Ω_1 and Ω_2. Each iteration step consists of two half steps, associated with the two subdomains

$$\begin{aligned} Lu_1^n &= f, & \text{in } \Omega_1 \ , \\ u_1^n &= g, & \text{on } L_0 := \partial\Omega \cap \partial\Omega_1 \ , \\ u_1^n &= u_2^{n-1}, & \text{on } L_2 := \partial\Omega_1 \setminus \partial\Omega \ , \end{aligned}$$

$$\begin{aligned} Lu_2^n &= f, & \text{in } \Omega_2 \ , \\ u_2^n &= g, & \text{on } L_3 := \partial\Omega \cap \partial\Omega_2 \ , \\ u_2^n &= u_1^n, & \text{on } L_1 := \partial\Omega_2 \setminus \partial\Omega \ . \end{aligned}$$

We remark that the Dirichlet boundary conditions on L_1 and L_2 are obtained by the solution on Ω_1 and Ω_2 in the previous half step, respectively. This classical Schwarz method can also be rewritten in a variational form, and the error propagation operator can easily be given in terms of two projection operators. We will not consider any further details and instead refer to [SBG96, QV99, Wid88, Wid99].

In the following, we consider only the variational formulation of general Schwarz methods. The starting point is the variational problem

$$a(u, v) = f(v), \quad v \in V \ , \tag{2.1}$$

where V is a finite element space, and the bilinear form $a(\cdot, \cdot)$ is associated with a selfadjoint, elliptic operator. Each Schwarz method is then based on a suitable decomposition of the finite dimensional space V into subspaces

$$V = V_0 + V_1 + \cdots V_N$$

and on projection-like operators T_i, $0 \le i \le N$, mapping V onto these subspaces. This decomposition does not have to be a direct sum. Typically, the subspaces are related to a sequence of nested triangulations or with basis functions having support in different subdomains. To obtain quasi-optimal results in the second case, very often, requires the use of a coarse global space. In the following two subsections, we briefly discuss the additive and multiplicative Schwarz method, and basic tools to establish bounds for the condition number and the error propagation.

2.1.1 Additive Schwarz Methods

The additive Schwarz variant is an important type of Schwarz methods. It provides a new operator equation which can be much better conditioned than the original discrete elliptic problem. Very often, the arising system can be solved efficiently by the conjugate gradient method. The quasi-projection operator $T_i : V \longrightarrow V_i$, is defined by means of an additional symmetric positive definite bilinear form $\tilde{a}_i(\cdot, \cdot)$ on $V_i \times V_i$

$$\tilde{a}_i(T_i w, v) := a(w, v), \quad v \in V_i .$$

In the case that we choose $\tilde{a}_i(\cdot, \cdot) = a(\cdot, \cdot)$, the operator T_i is the orthogonal projection onto V_i with respect to the bilinear form $a(\cdot, \cdot)$. The additive Schwarz operator is given by

$$T_{\text{add}} := \sum_{i=0}^{N} T_i ,$$

and the variational problem (2.1) can be rewritten as $T_{\text{add}} u = g$, where the right hand side g is defined as $g := \sum_{i=0}^{N} g_i$ with

$$\tilde{a}_i(g_i, v) := f(v), \quad v \in V_i .$$

The right hand side g is chosen so that the new problem has the same solution as the original one. Using the bilinear form $a(\cdot, \cdot)$ as inner product, we can apply the conjugate gradient method to the preconditioned problem. Then, the convergence rate in the energy norm can be bounded in terms of the condition number of T_{add}. An estimate for the smallest and largest eigenvalue of T_{add} is given by the following lemma; see [SBG96, Sect. 5.2].

Lemma 2.1. *Let us assume that for each $v \in V$ there exists a representation, $v = \sum_{i=0}^{N} v_i$, $v_i \in V_i$, such that*

$$\sum_{i=0}^{N} \tilde{a}_i(v_i, v_i) \leq C_0^2 \, a(v, v) . \tag{2.2}$$

Then, the operator T_{add} is invertible and a lower bound of $a(T_{\text{add}} v, v)$ is given by

$$C_0^{-2} a(v, v) \leq a(T_{\text{add}} v, v), \quad v \in V .$$

Moreover if the bilinear form $a(\cdot, \cdot)$ is bounded by a constant times $\tilde{a}_i(\cdot, \cdot)$ on the range of T_i, $0 \leq i \leq N$, i.e.,

$$a(T_i v, T_i v) \leq \omega \, \tilde{a}_i(T_i v, T_i v), \quad v \in V , \tag{2.3}$$

and if there exist constants $\epsilon_{ij} = \epsilon_{ji}$, $1 \leq i, j \leq N$, such that

$$a(v_i, v_j) \leq \epsilon_{ij} a(v_i, v_i)^{\frac{1}{2}} a(v_j, v_j)^{\frac{1}{2}}, \quad v_i \in V_i, \, v_j \in V_j , \tag{2.4}$$

then, an upper bound of $a(T_{\text{add}} v, v)$ can be given in terms of $a(v, v)$ and the spectral radius $\rho(\mathcal{E})$ of the matrix $\mathcal{E} := \{\epsilon_{ij}\}_{i,j=1}^{N}$:

$$a(T_{\text{add}} v, v) \leq \omega \, (1 + \rho(\mathcal{E})) \, a(v, v), \quad v \in V .$$

Proof. For convenience, we review the proof and refer to [SBG96, Sect. 5.2, Lemma 3] for a more detailed discussion. We start with the upper bound and consider the operator norm $\|T_i\|_a$

$$\|T_i\|_a^2 := \sup_{v \in V} \frac{a(T_i v, T_i v)}{a(v, v)} \ .$$

Using the definition of the operator T_i and assumption (2.3), we find

$$a(T_i v, T_i v) \le \omega \tilde{a}_i(T_i v, T_i v) = \omega a(v, T_i v) \le \omega a(v, v)^{\frac{1}{2}} a(T_i v, T_i v)^{\frac{1}{2}} \ ,$$

and thus $\|T_i\|_a \le \omega$. The upper bound of the norm $\|T_i\|_a$ and assumption (2.4) imply

$$a\Big(\sum_{i=1}^{N} T_i v, \sum_{i=1}^{N} T_i v \Big) = \sum_{i,j=1}^{N} a(T_i v, T_j v) \le \sum_{i,j=1}^{N} \epsilon_{ij} \, a(T_i v, T_i v)^{\frac{1}{2}} a(T_j v, T_j v)^{\frac{1}{2}}$$

$$\le \omega \sum_{i,j=1}^{N} \epsilon_{ij} a(v, T_i v)^{\frac{1}{2}} a(v, T_j v)^{\frac{1}{2}} \le \omega \rho(\mathcal{E}) \sum_{i=1}^{N} a(T_i v, v)$$

$$\le \omega \, \rho(\mathcal{E}) \, a(v, v)^{\frac{1}{2}} \, a\Big(\sum_{i=1}^{N} T_i v, \sum_{i=1}^{N} T_i v \Big)^{\frac{1}{2}} \ .$$

Adding T_0 completes the proof of the upper bound

$$a(T_{\text{add}} v, v) = a(T_0 v, v) + a\Big(\sum_{i=1}^{N} T_i v, v \Big) \le \omega(1 + \rho(\mathcal{E})) a(v, v) \ .$$

To prove the lower bound, we use the decomposition of assumption (2.2)

$$a(v, v) = \sum_{i=0}^{N} a(v, v_i) = \sum_{i=0}^{N} \tilde{a}_i(T_i v, v_i)$$

$$\le \sum_{i=0}^{N} \tilde{a}_i(T_i v, T_i v)^{\frac{1}{2}} \tilde{a}_i(v_i, v_i)^{\frac{1}{2}} \le \Big(\sum_{i=0}^{N} \tilde{a}_i(T_i v, T_i v) \Big)^{\frac{1}{2}} \Big(\sum_{i=0}^{N} \tilde{a}_i(v_i, v_i) \Big)^{\frac{1}{2}}$$

$$\le C_0 a(v, v)^{\frac{1}{2}} \Big(\sum_{i=0}^{N} a(T_i v, v) \Big)^{\frac{1}{2}} \le C_0 \, a(v, v)^{\frac{1}{2}} \, a(T_{\text{add}} v, v)^{\frac{1}{2}} \ .$$

$\square$

We remark that $\epsilon_{ij} \le 1$ and thus $\rho(\mathcal{E}) \le N$. Very often, an upper bound for $\rho(\mathcal{E})$ independent of the number of subspaces can be given. Thus, for many interesting applications, it is a routine matter to obtain a good bound for the largest eigenvalue of T_{add}. For multilevel techniques, a strengthened Cauchy–Schwarz inequality plays an important role, and very often we find $\epsilon_{ij} \le C q^{-|i-j|}$ with $q < 1$. Then, $\rho(\mathcal{E}) \le C$ is bounded independently of the number of subspaces; see, e.g., [Yse86].

In the case of overlapping domain decomposition methods, a coloring argument is used. Each subspace is associated with a color such that the subspaces V_i and V_j are a-orthogonal, i.e., $\epsilon_{ij} = 0$, if they have the same color. Then, the spectral radius $\rho(\mathcal{E})$ is bounded by the minimal number of required colors; see Fig. 2.4. All subspaces having the same color can be grouped together into a class of subspaces. The number of non zero entries in each row of $\mathcal{E}$ is bounded by the number of classes. Moreover, each class can be regarded for the analysis and the implementation as one subspace; see, e.g., [SBG96].

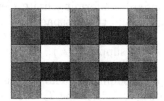

Fig. 2.4. Coloring of subdomains into four classes

If $\tilde{a}_i(\cdot, \cdot) = a(\cdot, \cdot)$ restricted on $V_i \times V_i$, (2.3) holds with $\omega = 1$. The delicate point in the analysis of an additive Schwarz method is often to find an optimal constant C_0 which measures the stability of the decomposition.

A classical example of an additive Schwarz method is the Jacobi method. Here, each subspace is one dimensional and the number of subspaces is equal the dimension of V. In the following, we briefly review three typical situations of decompositions of V into subspaces in the case of piecewise linear Lagrangian finite elements. Many Schwarz variants have been designed and analyzed for this standard case in 2D and in 3D. Recently, they have also been generalized to higher order elements including spectral and hp-methods [Cas97, GC97, GC98, Pav94a, Pav94b, PW97], elliptic systems [PW00a, PW00b], and vector field discretizations [HT00, Tos00, TWW00, WTW00].

Our first example is a two-level additive method with overlap. The domain Ω is connected with two triangulations, a macro-triangulation $\mathcal{T}_H$ and a fine triangulation $\mathcal{T}_h$. In addition, Ω is decomposed into overlapping subdomains Ω_i. For simplicity, Fig. 2.5 shows the special case where each subdomain is the union of elements of the fine triangulation.

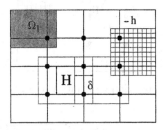

Fig. 2.5. Macro-triangulation and fine triangulation on Ω

Each subdomain does not have to be the union of elements in the fine triangulation, and the decomposition into subdomains and the macro-triangulation do not have to be related. We assume that the diameters of the elements of the macro-triangulation are roughly the same as the diameters of the adjacent subdomains, and that both, elements and subdomains, are shape regular. The minimal width of the overlapping region of two adjacent subdomains is denoted by δ, and we call δ/H the relative overlap. Then, $V_0 := V_H$ is just the standard conforming finite element space associated with the macro-triangulation, and the subspaces V_i, are given by

$$V_i := \{v \in V \mid \operatorname{supp} v \subset \overline{\Omega}_i\}, \quad 1 \le i \le N ,$$

where N is the number of subdomains. The degrees of freedom of V_0 are marked by filled circles in Fig. 2.5. Defining T_i as orthogonal projection with respect to the bilinear form $a(\cdot, \cdot)$, the additive Schwarz method yields a condition number bounded linearly by the inverse relative overlap

$$\kappa(T_{\mathrm{add}}) \le C\left(1 + \frac{H}{\delta}\right) ;$$

see [DW94, Wid99]. It has been shown only recently in [Bre00] that this estimate is sharp if the width of the overlap is bounded by ch. We remark that the results holds also for the 3D case.

The next example is an iterative substructuring method for P_1-Lagrangian finite elements; see [BPS86b, BPS89]. In [BPS89], a wirebasket algorithm in 3D has been proposed. Here, the condition number of the additive operator is bounded by a polylogarithmic factor. Three different types of subspaces define the operator T_{add}; see Fig. 2.6. A coarse subspace is necessary to obtain bounds for the condition number independent of the number of subdomains. As in the case of the overlapping variant, the coarse subspace is the finite element space associated with the macro-triangulation, i.e., $V_0 := V_H$.

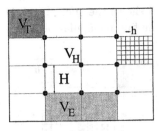

Fig. 2.6. Decomposition of V into three types of subspaces

In contrast to the overlapping case, we now consider a non-overlapping decomposition of Ω into subdomains. The elements T of the macro-triangulation define the subdomains, and as before we set

$$V_T := \{v \in V \mid \text{supp } v \subset \overline{T}\}, \quad T \in \mathcal{T}_H .$$

Since the coarse elements are non-overlapping the sum over the V_T is direct and a proper subspace of V. We add a third type of subspaces, which are associated with the individual edges of the coarse triangulation

$$V_E := \{v \in V \mid \text{supp } v \subset \overline{T}_1 \cup \overline{T}_2, a(v, w_T) = 0, w_T \in V_T, T \in \mathcal{T}_H\} ,$$

where $\partial T_1 \cap \partial T_2 = E \in \mathcal{E}_H$, and $\mathcal{E}_H$ is the set of edges of the macro-triangulation. The dimension of V_E is given by the number of fine edges $e \subset E$ minus one. Each element in V_E is uniquely defined by its values at the vertices on the macro-edge E, and is obtained by a discrete harmonic extension. Using the exact projections onto these subspaces, the iterative substructuring method provides a polylogarithmic upper bound for the condition number of the additive Schwarz method of the form

$$\kappa(T_{\text{add}}) \leq C\Big(1 + (\log \frac{H}{h})^2\Big) ;$$

see [BPS86a, BPS86b]. Recently, it has been shown in [BS00] that this bound is sharp. In the case of P_1-Lagrangian finite elements, there is an essential difference between the 2D and the 3D case. To obtain quasi-optimal results for the 3D case, the coarse space V_0 has to be modified in the case of highly discontinuous coefficients. The projection onto the finite element space V_H is replaced by a different suitable low dimensional problem. One of the successful methods is based on a wirebasket type space and a quasi-projection is selected; see [BPS89]. We refer to [DSW94], for different possibles choices of V_0 and T_0. In Subsect. 2.2.2, we establish an iterative substructuring method for Raviart–Thomas vector fields in 3D. As we will see, there is a principal difference between the results for this finite element space and that for the standard Lagrangian finite element space.

Our last example is the well known hierarchical basis method in 2D introduced in [Yse86]. Here, the subspaces are not associated with geometrical subdomains but with a nested sequence of triangulations, $\mathcal{T}_l$, $0 \leq l \leq N$. The triangulation $\mathcal{T}_{l+1}$ is obtained from $\mathcal{T}_l$ by decomposing each element into four congruent subelements. Now, the subspace V_i is associated with the triangulation $\mathcal{T}_i$ and defined by

$$V_i := V_{\mathcal{T}_i} \setminus V_{\mathcal{T}_{i-1}}, \quad 0 \leq i \leq N ,$$

where $V_{\mathcal{T}_l}$, $0 \leq l \leq N$, is the standard finite element space associated with the triangulation $\mathcal{T}_l$, and $V_{\mathcal{T}_{-1}} := \emptyset$. Using this notation, we have $V = V_{\mathcal{T}_N}$. In contrast to the first example, these spaces form a direct sum. The filled circles in Fig. 2.7 illustrate the nodes of the finite element space V. Each circle having the same size is a degree of freedom of the same V_i, the larger the diameter is the smaller the index i. The largest circles belong to V_0 and the smallest to V_N.

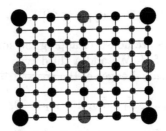

Fig. 2.7. Hierarchical decomposition of the nodes of V

The implementation can be based on a quasi-projection. One possibility is to define the modified bilinear form $\tilde{a}(\cdot,\cdot)$ by

$$\tilde{a}(v_i, w_i) := \sum_j \alpha_j \, \beta_j \, a(\phi_j^i, \phi_j^i) \ ,$$

where ϕ_j^i are the hierarchical nodal basis functions of V_i, and $v_i := \sum_j \alpha_j \phi_j^i$, $w_i := \sum_j \beta_j \phi_j^i$. Then, a straightforward computation shows that the quasi-projection T_i is given by

$$T_i v = \sum_j \frac{a(v, \phi_j^i)}{a(\phi_j^i, \phi_j^i)} \, \phi_j^i \ .$$

This can be also interpreted as a decomposition of V_i into one dimensional subspaces associated with the nodal basis functions ϕ_j^i

$$V_i = \sum_j \text{span} \ \phi_j^i =: \sum_j V_j^i \ .$$

In this case, the exact projections P_j^i onto the one dimensional spaces V_j^i are given by

$$P_j^i v = \frac{a(v, \phi_j^i)}{a(\phi_j^i, \phi_j^i)} \, \phi_j^i \ ,$$

and we find $T_i = \sum_j P_j^i$. The following quasi-optimal result holds for the 2D case

$$\kappa(T_{\text{add}}) \leq C(1 + N^2) \ ;$$

see [Yse86, Ban96], where N is the number of refinement levels. In the 3D case, only an exponential bound in N can be obtained.

For all three examples, the upper bound for the norm of T_{add} is bounded independently of the number of subdomains. A coloring argument is used to bound the spectral radius of $\mathcal{E}$ in Lemma 2.1 for the two first examples. In the case of the hierarchical decomposition, the constant upper bound is based on a strengthened Cauchy–Schwarz inequality. It can be shown that

$$a(v_i, v_j) \leq C\, 2^{-|i-j|} a(v_i, v_i)^{\frac{1}{2}} a(v_j, v_j)^{\frac{1}{2}}, \quad v_i \in V_i,\ v_j \in V_j \ ;$$

see, e.g., [Yse86]. In Subsect. 2.2.3, we show that the same qualitative bound can be obtained for Raviart–Thomas finite elements even in the 3D case.

2.1.2 Multiplicative Schwarz Methods

The second important family of Schwarz methods is the multiplicative one. A theoretical analysis of this type can be found in [BPWX91b, CW92, CW93, SBG96, Wid99]. Examples are given by multigrid methods; see Subsect. 2.1.3.

The multiplicative Schwarz variant is defined by

$$T_{\mathrm{mul}} := \mathrm{Id} - \prod_{i=0}^{N} (\mathrm{Id} - T_i) \ .$$

In contrast to the additive type, the definition depends on the ordering of the subspaces. A classical example of a multiplicative Schwarz method is the well known Gauß–Seidel method.

Using the multiplicative variant as an iteration method, the error propagation operator can be written as

$$E_{\mathrm{mul}} = \prod_{i=0}^{N} (\mathrm{Id} - T_i) \ .$$

The following lemma gives an upper bound for the norm of E_{mul}, a proof of which or quite similar results can be found in [BPWX91a, SBG96, Wid99].

Lemma 2.2. *Under the assumptions of Lemma 2.1 and $\omega < 2$, the energy norm of the error propagation operator of the multiplicative Schwarz method is bounded by*

$$\|E_{\mathrm{mult}}\|_a \leq \sqrt{1 - \frac{2 - \widetilde{\omega}}{(1 + 2\widetilde{\omega}^2 \rho(\mathcal{E})^2) C_0^2}} \ ,$$

where $\widetilde{\omega} := \max(1, \omega)$.

Proof. To obtain an upper bound for the norm of the error propagation operator E_{mul}, we set $E_i := (\mathrm{Id} - T_i) \ldots (\mathrm{Id} - T_0)$, $0 \leq i \leq N$, and $E_{-1} := \mathrm{Id}$ and define R_i by $R_i := (2 - T_i) T_i$. Then, we have the identity

$$E_{i-1}^T R_i E_{i-1} = E_{i-1}^T E_{i-1} - E_i^T E_i, \quad 0 \leq i \leq N \ .$$

Here, the transpose is taken with respect to the bilinear form $a(\cdot, \cdot)$. Summing over all i and using a telescopic cancellation yield

$$E_{\mathrm{mul}}^T E_{\mathrm{mul}} = \mathrm{Id} - \sum_{i=0}^{N} E_{i-1}^T R_i E_{i-1} \ .$$

Under the assumption $\omega < 2$, the operators R_i are symmetric and positive definite. Moreover by means of $\|T_i\|_a \leq \omega$, we find $R_i \geq (2 - \widetilde{\omega})T_i$ and thus

$$E_{\text{mul}}^T E_{\text{mul}} \leq \text{Id} - (2 - \widetilde{\omega}) \sum_{i=0}^{N} E_{i-1}^T T_i E_{i-1} \ . \tag{2.5}$$

Using the identity $E_j + T_j E_{j-1} = E_{j-1}$, $0 \leq j \leq N$, we obtain by summing over j and canceling common terms

$$\text{Id} = E_{i-1} + \sum_{j=0}^{i-1} T_j E_{j-1}, \quad 1 \leq i \leq N \ .$$

Now, this identity can be used to obtain an upper bound for $a(T_i v, v)$. The assumption (2.4) and the bound for $\|T_i\|_a$ together with the Schwarz inequality yield

$$a(T_i v, v) = a(T_i v, E_{i-1} v) + a(T_i v, T_0 v) + \sum_{j=1}^{i-1} a(T_i v, T_j E_{j-1} v)$$

$$\leq a(T_i v, v)^{\frac{1}{2}} \left(a(T_i T_0 v, T_0 v)^{\frac{1}{2}} + \widetilde{\omega} \sum_{j=1}^{i} \epsilon_{ij} a(T_j E_{j-1} v, E_{j-1} v)^{\frac{1}{2}} \right)$$

$$\leq a(T_i v, v)^{\frac{1}{2}} \left(a(T_i T_0 v, T_0 v)^{\frac{1}{2}} + \widetilde{\omega} \sum_{j=1}^{N} \epsilon_{ij} a(T_j E_{j-1} v, E_{j-1} v)^{\frac{1}{2}} \right)$$

$$a(T_i v, v) \leq 2 a(T_i T_0 v, T_0 v) + 2\widetilde{\omega}^2 \left(\sum_{j=1}^{N} \epsilon_{ij} a(T_j E_{j-1} v, E_{j-1} v)^{\frac{1}{2}} \right)^2 \ .$$

Here, we have used that $\epsilon_{ii} = 1$, $\epsilon_{ij} \geq 0$, and the definition of $\widetilde{\omega}$. Finally, we sum the last inequality from $i = 1$ to N and use $a((T_{\text{add}} - T_0)T_0 v, T_0 v) \leq \omega \, \rho(\mathcal{E}) \, a(T_0 v, T_0 v)$. Then, $\rho(\mathcal{E}) \geq 1$ gives

$$a(T_{\text{add}} v, v) = a((T_{\text{add}} - T_0)v, v) + a(T_0 v, v)$$

$$\leq (1 + 2\widetilde{\omega}^2 \rho(\mathcal{E})) a(T_0 v, v) + 2\widetilde{\omega}^2 \rho(\mathcal{E})^2 \sum_{j=1}^{N} a(T_j E_{j-1} v, E_{j-1} v)$$

$$\leq (1 + 2\widetilde{\omega}^2 \rho(\mathcal{E})^2) \sum_{j=0}^{N} a(T_j E_{j-1} v, E_{j-1} v) \ .$$

Now in terms of (2.5) and the lower bound of Lemma 2.1, the norm of the error propagation operator of the multiplicative Schwarz variant can be bounded by

$$\|E_{\text{mult}}\|_a^2 \leq 1 - \frac{2 - \widetilde{\omega}}{1 + 2\widetilde{\omega}^2 \rho(\mathcal{E})^2} \inf_{v \in V} \frac{a(T_{\text{add}} v, v)}{a(v, v)} \leq 1 - \frac{2 - \widetilde{\omega}}{(1 + 2\widetilde{\omega}^2 \rho(\mathcal{E})^2)C_0^2} \ .$$

$\square$

Obviously, the bound is of interest only if $\omega < 2$. If we use the exact projection for T_i, $\omega = 1$. Otherwise a simple rescaling of the projection-like operators T_i

will always yield $\omega < 2$. However, a rescaling of some T_i also influences the constant C_0 in Lemma 2.1.

We note that the multiplicative variant of a Schwarz method, very often, results in fewer iterations to obtain a given accuracy than the corresponding additive method. However, additive methods can be very often more easily parallelized. In contrast to the additive variant, the multiplicative one is in general non-symmetric. A symmetrized form can be defined by $T_{\mathrm{mul}} + T_{\mathrm{mul}}^T$, where T_{mul}^T is the adjoint operator or by

$$\mathrm{Id} - (\mathrm{Id} - T_N)\ldots(\mathrm{Id} - T_1)(\mathrm{Id} - T_0)^2(\mathrm{Id} - T_1)\ldots(\mathrm{Id} - T_N) \ .$$

In the case that the exact projection is used for T_0, we can use $(\mathrm{Id} - T_0)^2 = (\mathrm{Id} - T_0)$ to eliminate one step. Many other variants, such as hybrid schemes, can be obtained by combining additive and multiplicative components.

As an example of a multiplicative Schwarz variant, we mention the $\mathcal{V}$-cycle for standard Lagrangian finite elements [Bra93, Osw94]. The corresponding additive variant is the well known BPX-preconditioner [BPX90a]. We refer to [GO95] for an abstract theory for additive and multiplicative Schwarz methods.

A different way to analyze a multigrid method, in particular $\mathcal{W}$-cycles, is discussed in the following; we will give a brief introduction to multigrid methods in Subsect. 2.1.3. In Sect. 2.3 and Sect. 2.5, we focus on two multigrid methods for mortar finite elements. To prove level independent convergence rates, we do not use the Schwarz theory but instead ideas discussed in [Hac85].

2.1.3 Multigrid Methods

Multigrid methods provide optimal iteration schemes for the solution of systems of equations arising from the discretization of elliptic boundary value problems. The computational cost to reach a given accuracy is linear in the number of unknowns. For a general introduction to these methods and related techniques, we refer to [BH83, Bra93, BS94, BY93, Hac85, McC87, Osw94]. There are two approaches of establishing bounds for the convergence rate of a multigrid method. One uses the abstract framework of multiplicative Schwarz methods and can be found for example in [BS94, Bra93]. We do not discuss this approach here. In our analysis, we instead follow the arguments in [Hac85] and establish suitable approximation and smoothing properties; a level independent convergence rate for the $\mathcal{W}$-cycle can then be shown only if the number of smoothing steps is large enough.

The additive variant of a Schwarz method defines, in general, a preconditioned system and Krylov subspace methods, e.g., a conjugate gradient method, can be used as accelerators. In contrast to this, the multiplicative form itself provides an efficient iterative solver. Here, we briefly discuss the basic ingredients of a multigrid method. It consists of smoothing steps and correction steps in lower dimensional finite element spaces. The smoothing part damps the oscillatory component of the error whereas the correction is associated with a problem on a coarser mesh. Multigrid techniques are motivated by the following observation: Classical iteration schemes restricted to the subspace of high frequency show a fast convergence. The subspace of low frequencies cause the observed slow convergence. Figure 2.8 illustrates the smoothing effect of a Gauß–Seidel method. After a few smoothing steps, the error on the fine mesh can be accurately represented on a coarser mesh.

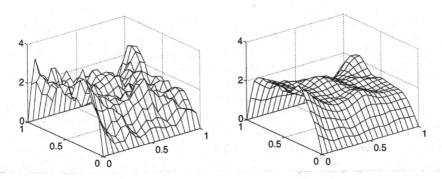

Fig. 2.8. Effect of one symmetric Gauß–Seidel smoothing step

Similar to the multilevel method described in Subsect. 2.1.1, a multigrid method is often associated with a nested sequence of triangulations and corresponding finite element spaces V_k, $0 \le k \le j$. We restrict ourselves to the case of nested spaces, $V_0 \subset V_1 \dots \subset V_j$, and refer to [BDH99b, Bre89, BV90, Osw94] for a discussion of nonconforming situations.

The iterative solver for $A_l u_l = f_l$, on level l, is defined recursively. It depends on the choice of the transfer operators $I_l^{l-1} : V_l \longrightarrow V_{l-1}$, and $I_{l-1}^l : V_{l-1} \longrightarrow V_l$, $1 \le l \le j$, and the smoothing operators G_1, G_2

$$u_l^{\nu+1} := MG(l, u_l^\nu, f_l), \quad \nu \ge 0 ,$$

where u_l^0 is the initial guess. On level zero, $MG(0, u_0^\nu, f_0)$ is defined by $MG(0, u_0^\nu, f_0) := A_0^{-1} f_0$. The multigrid operator $MG(l, u_l^\nu, f_l)$, $l > 0$, is given recursively and consists of three steps – presmoothing, error correction and postsmoothing. The correction step depends on the transfer operator.

- Presmoothing step: For $1 \le i \le m_1$, let

$$z_l^i := z_l^{i-1} + G_1^{-1}(f_l - A_l z_l^{i-1}) , \qquad (2.6)$$

where $m_1 \ge 1$, G_1 is a suitable smoothing operator and $z_l^0 := u_l^\nu$.
- Correction step: $d_l := f_l - A_l z_l^{m_1}$
  - Defect restriction: $d_{l-1} := I_l^{l-1} d_l$

– Coarse grid correction: For $1 \leq j \leq p$, let

$$q_{l-1}^{j} := MG(l-1, q_{l-1}^{j-1}, d_{l-1}) \ ,$$

where $q_{l-1}^{0} := 0$.

– Prolongation of the correction: $z_{l}^{m_1+1} := z_{l}^{m_1} + I_{l-1}^{l} q_{l-1}^{p}$
• Postsmoothing step: For $m_1 + 2 \leq i \leq m_1 + 1 + m_2$, let

$$z_{l}^{i} := z_{l}^{i-1} + G_{2}^{-1}(f_l - A_l z_{l}^{i-1}) \ ,$$

where $m_2 \geq 0$ and G_2 is a suitable smoothing operator. Finally, a multigrid step is defined by

$$MG(l, u_{l}^{\nu}, f_l) := z_{l}^{m_1+m_2+1} \ .$$

The case $p = 1$ is called the $\mathcal{V}$-cycle and the case $p = 2$ the $\mathcal{W}$-cycle. In practice, the restriction I_{l}^{l-1} is, very often, chosen as transpose of the prolongation I_{l-1}^{l}, and in the case of nested spaces $V_{l-1} \subset V_l$ the natural injection is taken for I_{l-1}^{l}. The easiest type of a smoother is a damped Richardson method where the damping factor has to be smaller than one over the largest eigenvalue of A_l. Symmetric Gauß–Seidel methods and incomplete LU-factorizations provide more robust smoothers. The choice $m_1 = m_2$, $G_1 = G_2^T$, and $I_{l-1}^{l} = (I_{l-1}^{l})^T$ yields a symmetric iterative solution scheme.

One possibility to prove level independent convergence rates for $\mathcal{W}$-cycles relies on suitable approximation and smoothing properties [Bra97, Hac85]. The analysis of the multigrid case is based on the two grid case and a perturbation argument. In the two grid case, approximation and smoothing properties yield that the convergence rates are independent of the meshsize provided that the number of smoothing steps is large enough. Additionally, a stability estimate for the smoothing scheme is required for the multigrid analysis.

2.2 Vector Field Discretizations

In this section, we consider a boundary value problem for vector fields, associated with the divergence operator. We focus on the 3D case and the well known Raviart–Thomas finite elements; see, e.g., [BF91]. Applications of these vector fields can be found in [AFW97]. In Subsect. 2.2.2, we develop an iterative substructuring method for this class of finite elements and in Subsect. 2.2.3, one based on a hierarchical basis. We note that there are several interesting differences when compared to the H^1-case.

We consider the following second order partial differential equation for vector fields

$$\begin{aligned} L\mathbf{u} &:= -\mathbf{grad}\,(a\,\mathrm{div}\,\mathbf{u}) + B\,\mathbf{u} = \mathbf{f}, \quad &\text{in } \Omega \ , \\ \mathbf{u} \cdot \mathbf{n} &= 0, \quad &\text{on } \partial\Omega \ , \end{aligned} \qquad (2.7)$$

where Ω is a bounded polyhedral domain in $\mathbb{R}^3$ of unit diameter, and $\mathbf{n}$ its outward normal. We assume that $\mathbf{f} \in (L^2(\Omega))^3$, that the coefficient matrix B is a symmetric uniformly positive matrix-valued function with $b_{ij} \in L^\infty(\Omega)$, $1 \le i, j \le 3$, and that $a \in L^\infty(\Omega)$ satisfies $a \ge a_0 > 0$ almost everywhere.

The weak formulation of problem (2.7) is defined in a vector valued Hilbert space of which $(H^1(\Omega))^3$ is a proper subspace. Applying Green's formula, we find that $H(\text{div};\Omega)$ is a suitable space

$$H(\text{div};\Omega) := \left\{ \mathbf{v} \in (L^2(\Omega))^3 \mid \; \text{div}\,\mathbf{v} \in L^2(\Omega) \right\} \; ,$$

equipped with the inner product $(\cdot, \cdot)_{\text{div}}$

$$(\mathbf{w}, \mathbf{v})_{\text{div}} := \int_\Omega \mathbf{w} \cdot \mathbf{v} \; dx + \int_\Omega \text{div}\,\mathbf{w} \, \text{div}\,\mathbf{v} \; dx \; .$$

The corresponding norm is defined by $\|\mathbf{w}\|_{\text{div}}^2 := (\mathbf{w}, \mathbf{w})_{\text{div}}$. The boundary condition in (2.7) will be imposed on the space. A trace theorem given in [BF91] shows that the normal component, $\mathbf{v} \cdot \mathbf{n}$, of any vector $\mathbf{v} \in H(\text{div};\Omega)$, on the boundary $\partial\Omega$, belongs to the space $H^{-1/2}(\partial\Omega)$. The subspace of vectors in $H(\text{div};\Omega)$ with vanishing normal component on $\partial\Omega$ is called $H_0(\text{div};\Omega)$.

We are now prepared to formulate the weak variational form of the boundary value problem (2.7): Find $\mathbf{u} \in H_0(\text{div};\Omega)$ such that

$$a(\mathbf{u}, \mathbf{v}) = \int_\Omega \mathbf{f} \cdot \mathbf{v} \; dx, \quad \mathbf{v} \in H_0(\text{div};\Omega) \; , \tag{2.8}$$

where the bilinear form $a(\cdot, \cdot)$ is given by

$$a(\mathbf{w}, \mathbf{v}) := \int_\Omega (a \, \text{div}\,\mathbf{w} \, \text{div}\,\mathbf{v} + B \, \mathbf{w} \cdot \mathbf{v}) \; dx, \quad \mathbf{w}, \mathbf{v} \in H(\text{div};\Omega) \; .$$

An energy norm defined by $\|\mathbf{w}\|^2 := a(\mathbf{w}, \mathbf{w})$ is associated with the bilinear form $a(\cdot, \cdot)$. In particular, this energy norm is equivalent to the Hilbert space norm $\| \cdot \|_{\text{div}}$. The equivalence constants depend on the coefficients and can be quite large or quite small.

Before defining the Raviart–Thomas space of lowest order, we consider the trace of an element in $H(\text{div};\Omega)$ in more detail. Throughout the rest of this section, we will work with scaled Sobolev norms; the size of the domains is reflected in weight factors. Given a bounded open Lipschitz domain $D \subset \Omega$, with a boundary ∂D and a diameter H_D, let $| \cdot |_{s;D}$ denote the semi norm of the Sobolev space $H^s(D)$. Then the scaled norms are defined by

$$\|\phi\|_{1;D}^2 = |\phi|_{1;D}^2 + \frac{1}{H_D{}^2}\|\phi\|_{0;D}^2, \quad \phi \in H^1(D) \; ,$$

and

$$\|\phi\|^2_{\frac{1}{2};\partial D} = |\phi|^2_{\frac{1}{2};\partial D} + \frac{1}{H_D}\|\phi\|^2_{0;\partial D}, \quad \phi \in H^{\frac{1}{2}}(\partial D) \ .$$

In the case that $D = \Omega$, we will drop the reference to the region. In contrast to the previous sections, the $H^{-1/2}$-norm will from now on reflect this scaling, and it is defined by

$$\|\mathbf{w} \cdot \mathbf{n}\|_{-\frac{1}{2};\partial D} := \sup_{\substack{\phi \in H^{\frac{1}{2}}(\partial D) \\ \phi \neq 0}} \frac{\langle \mathbf{w} \cdot \mathbf{n}, \phi \rangle}{\|\phi\|_{\frac{1}{2};\partial D}}, \quad \mathbf{w} \in H(\mathrm{div}\,;\Omega) \ ,$$

where $\langle \cdot, \cdot \rangle$ represents the duality pairing between $H^{-1/2}(\partial D)$ and $H^{1/2}(\partial D)$. The following lemma provides a basic estimate for the trace of an element in $H(\mathrm{div}\,;\Omega)$. The proof is quite elementary and is based on Green's formula and a scaling argument.

Lemma 2.3. *There exists a constant C, which is independent of the diameter of D but depend on the shape regularity of D, such that for $\mathbf{w} \in H(\mathrm{div}\,;\Omega)$*

$$\|\mathbf{w} \cdot \mathbf{n}\|^2_{-\frac{1}{2};\partial D} \leq C \left(\|\mathbf{w}\|^2_{0;D} + H^2_D \|\mathrm{div}\,\mathbf{w}\|^2_{0;D} \right) \ .$$

We will use the well known Raviart–Thomas spaces for the discretization of (2.8). The finite element approximation is given on a triangulation $\mathcal{T}_h$, the elements of which are denoted by T. The set of interior faces and edges of the triangulations $\mathcal{T}_h$ is called $\mathcal{F}_h$ and $\mathcal{E}_h$, respectively.

2.2.1 Raviart–Thomas Finite Elements

Our study concerns the lowest order Raviart–Thomas finite elements for the discrete approximation of (2.7); see [BF91]. In Sect. 1.4, this space was already used to define a mortar finite element discretization. Here, we work in 3D and the definition of the degrees of freedom is essential, and we therefore consider this space in more detail in the rest of this subsection.

The global Raviart–Thomas finite element space $\mathcal{RT}(\Omega;\mathcal{T}_h)$ is defined by means of the local ones

$$\mathcal{RT}(\Omega;\mathcal{T}_h) := \left\{ \mathbf{w} \in H(\mathrm{div}\,;\Omega)| \quad \mathbf{w}_{|_T} \in \mathcal{RT}(T), T \in \mathcal{T}_h \right\} \ ,$$

where $\mathcal{RT}(T)$ stands for the local Raviart–Thomas space. In the case of a hexahedral triangulation, where the elements are cubes, the local space has dimension six, while in the case of a simplicial triangulation, where the elements are tetrahedras, the local space is four dimensional. For a cube with sides parallel to the coordinate axes, $\mathcal{RT}(T)$ is given by

$$\mathcal{RT}(T) := \begin{pmatrix} \alpha_1 + \beta_1 x \\ \alpha_2 + \beta_2 y \\ \alpha_3 + \beta_3 z \end{pmatrix}, \quad \alpha_i, \beta_i \in \mathbb{R} \ .$$

In the case of a tetrahedra, $\mathcal{RT}(T)$ is defined by

$$\mathcal{RT}(T) := \begin{pmatrix} \alpha_1 + \beta x \\ \alpha_2 + \beta y \\ \alpha_3 + \beta z \end{pmatrix}, \quad \alpha_i, \beta \in \mathbb{R} . \tag{2.9}$$

The degrees of freedom of $\mathcal{RT}(\Omega; \mathcal{T}_h)$ are given by the averages of the normal components over the faces F of the triangulation:

$$\lambda_F(\mathbf{w}) := \frac{1}{|F|} \int_F \mathbf{w} \cdot \mathbf{n} \, d\sigma .$$

Here, $|F|$ is the area of the face F and the direction of the normal can be fixed arbitrarily for each face. This formula also defines the natural interpolation operator from $H(\text{div}; \Omega)$ onto the space $\mathcal{RT}(\Omega; \mathcal{T}_h)$. We note that the normal component of any Raviart–Thomas vector field is constant and continuous across each face, see Fig. 2.9.

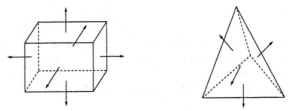

Fig. 2.9. Local degrees of freedom of a lowest order Raviart–Thomas vector field

We define the subspace of vectors with vanishing normal components on the boundary of Ω by

$$V_h := \mathcal{RT}(\Omega; \mathcal{T}_h) \cap H_0(\text{div}; \Omega) .$$

As in the case of Lagrangian finite elements, the L^2-norm of these discrete vector fields can be bounded from above and below by means of the values of their degrees of freedom. For $\mathbf{w} \in \mathcal{RT}(T)$, we have

$$c \sum_{F \subset \partial T} \left(H_F^{\frac{3}{2}} \lambda_F(\mathbf{w}) \right)^2 \leq \|\mathbf{w}\|_{0;T}^2 \leq C \sum_{F \subset \partial T} \left(H_F^{\frac{3}{2}} \lambda_F(\mathbf{w}) \right)^2 , \tag{2.10}$$

where H_F is the diameter of the face F. The equivalence constants depend on the aspect ratio of the elements, but do not depend on the diameter. Moreover, the following inverse estimate holds:

$$\|\text{div}\,\mathbf{w}\|_{0;T} \leq C \frac{1}{H_T} \|\mathbf{w}\|_{0;T}, \quad \mathbf{w} \in \mathcal{RT}(T) .$$

These bounds can easily be shown by using the affine equivalence of the elements of the triangulations and the finite dimension of the local spaces.

We note that relatively few studies exist of domain decomposition methods for $H(\text{div}\,;\,\Omega)$ and $H(\text{curl}\,;\,\Omega)$ in 3D; we refer to [CPRY97, HT00, Tos00], for two-level overlapping methods, to [AFW00, BDH$^+$99a, Hip96, Hip97, Hip98] for multilevel and multigrid methods, and to [AV99] for a study of an iterative substructuring method in $H(\text{curl}\,;\,\Omega)$. We refer to [AFW97, Bre92, EW92, Mat93a, Mat93b] and to the references therein, for some Schwarz methods for problems in $H(\text{div}\,;\,\Omega)$ in 2D. A certain class of multilevel methods for the mixed approximation of the Laplace equation is discussed in [Sar94]. Here, we construct a quasi-optimal iterative substructuring method as well as a hierarchical basis method for the 3D case.

2.2.2 An Iterative Substructuring Method

In this subsection, we introduce and analyze our iterative substructuring method, originally described in [WTW00]. For a quite general introduction to domain decomposition methods for vector fields discretizations, we refer to [Tos99].

We restrict ourselves to the case of a hexahedral triangulation $\mathcal{T}_h$ which is obtained by quasi-uniform refinement from a coarse macro-triangulation $\mathcal{T}_H$. From now on the generic elements, faces and edges of $\mathcal{T}_h$ are denoted by t, f, and e, and those of $\mathcal{T}_H$ by T, F, and E, respectively. The set of interior faces of the macro-triangulation $\mathcal{T}_H$ is called $\mathcal{F}_H$. We use a simplified notation for the ratio of the meshsizes between macro and fine triangulation. In the context of elementwise estimates, H/h denotes the ratio of the local meshsizes whereas in global bounds, H/h stands for the maximum of the local ratios. As we have seen in Sect. 2.1, the first step towards the introduction of an additive Schwarz method is to define a set of subspaces; see Lemma 2.1. As in the 2D case for standard Lagrangian finite elements, we introduce three different types of subspaces called V_H, V_F and V_T.

To obtain scalable bounds, we cannot avoid the use of a global space. But in contrast to the standard Lagrangian finite elements in 3D, the low dimensional Raviart–Thomas space associated with the macro-triangulation, $V_H := \mathcal{RT}(\Omega;\mathcal{T}_H) \cap H_0(\text{div}\,;\,\Omega)$, can be used for this purpose while at the same time the constant in the estimate of the condition number does not depend on the jumps of the coefficients.

The local spaces V_T, $T \in \mathcal{T}_H$, are associated with the elements, also called substructures, of the macro-triangulation. Each element in V_T,

$$V_T := \{\mathbf{v} \in V_h \mid \text{supp } \mathbf{v} \subset \overline{T}\}, \quad T \in \mathcal{T}_H \ ,$$

has support contained in $\overline{T}$. Thus, the projections onto V_T are orthogonal.

The third type of subspaces is associated with the faces $F \in \mathcal{F}_H$ of the macro-triangulation. The supports of its elements are contained in two substructures T_1 and T_2. For each interior face $F \in \mathcal{F}_H$, there are two elements

T_1 and $T_2 \in \mathcal{T}_H$ such that $\overline{F} = \partial T_1 \cap \partial T_2$, and we set $\overline{T}_F := \overline{T}_1 \cup \overline{T}_2$. The face spaces are defined by

$$V_F := \{\mathbf{v} \in V_h \mid a(\mathbf{v}, \mathbf{w}) = 0, \; \mathbf{w} \in V_{T_1} + V_{T_2}, \; \mathrm{supp}\,\mathbf{v} \subset \overline{T}_F\} \; .$$

We note that an element $\mathbf{v} \in V_F$ is defined uniquely by its values $\mathbf{v} \cdot \mathbf{n}$ on F.

Now, we decompose V_h into the coarse space V_H, the face spaces V_F, $F \in \mathcal{F}_H$, and the interior spaces V_T, $T \in \mathcal{T}_H$, and we observe that the coarse space V_H is contained in the union of the face and interior spaces. The decomposition,

$$V_h = V_H + \sum_{T \in \mathcal{T}_H} V_T + \sum_{F \in \mathcal{F}_H} V_F \; , \tag{2.11}$$

on which our iterative substructuring method is based, is therefore not a direct sum. Figure 2.10 illustrates the local decomposition of V_h into the three different types of subspaces. Each type of subspace is symbolized by the faces which are associated with the degrees of freedom.

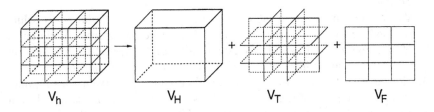

V_h V_H V_T V_F

Fig. 2.10. Decomposition of V_h into three types of subspaces

For simplicity, we restrict ourselves to the case that the exact projections onto the subspaces are used to define the additive Schwarz method. According to Lemma 2.1, the proof of an upper bound for the norm of T_{add} is elementary. It is based on a coloring argument, and the coefficients a and B do not enter into the bound. The crucial part is to find a decomposition of $\mathbf{v} \in V_h$ such that (2.2) holds with a C_0 as small as possible. A bound for the corresponding multiplicative variant follows from Lemma 2.2.

Observing that $V_h = \sum_{T \in \mathcal{T}_H} V_T + \sum_{F \in \mathcal{F}_H} V_F$ is a direct sum, the decomposition

$$\mathbf{v} = \mathbf{v}_H + \sum_{T \in \mathcal{T}_H} \mathbf{v}_T + \sum_{F \in \mathcal{F}_H} \mathbf{v}_F \; ,$$

with $\mathbf{v}_H \in V_H$, $\mathbf{v}_F \in V_F$, and $\mathbf{v}_T \in V_T$, is unique as soon as $\mathbf{v}_H$ has been fixed. A first step towards the proof of a sharp bound for C_0 is to find a suitable $\mathbf{v}_H \in V_H$. In the following subsubsection, we consider a standard interpolant ρ_H onto the global coarse subspace. By means of the discrete norm equivalence (2.10), we will establish stability bounds in the L^2- and the $H(\mathrm{div}\,;\Omega)$-norm. A detailed discussion of this operator can be found in [WTW00].

2.2.2.1 An Interpolation Operator onto V_H. An important role in the analysis of our additive Schwarz method is played by the interpolation ρ_H onto the global coarse subspace V_H. This interpolation operator is defined in terms of the degrees of freedom of V_H by

$$\lambda_F(\rho_H\mathbf{v}) := \frac{1}{|F|}\int\limits_F \mathbf{v}\cdot\mathbf{n}\,d\sigma, \quad F\in\mathcal{F}_H \ .$$

We note that the stability estimates for the interpolant ρ_H will enter into our estimate of the constant C_0. As we will see, ρ_H is not uniformly stable in h. The following lemma can be found in [WTW00] and gives a bound for the interpolant ρ_H in terms of the ratio H/h.

Lemma 2.4. *There exists a constant C, which depends only on the aspect ratios of $T\in\mathcal{T}_H$ and the elements of $\mathcal{T}_h$, such that for all $\mathbf{v}\in V_h$,*

$$\|\mathrm{div}\,(\rho_H\mathbf{v})\|^2_{0;T} \le \|\mathrm{div}\,\mathbf{v}\|^2_{0;T} \ , \tag{2.12}$$

$$\|\rho_H\mathbf{v}\|^2_{0;T} \le C\left(\left(1+\log\frac{H}{h}\right)\|\mathbf{v}\|^2_{0;T} + H_T^2\|\mathrm{div}\,\mathbf{v}\|^2_{0;T}\right) \ . \tag{2.13}$$

Proof. For a better understanding of the techniques, we review the proof given in [WTW00]. By a simple computation and the use of Green's formula, we find that $(\mathrm{div}\,(\rho_H\mathbf{v}))_{|T}$ is constant and

$$(\mathrm{div}\,(\rho_H\mathbf{v}))_{|T} = (\Pi_H(\mathrm{div}\,\mathbf{v}))_{|T} \ ,$$

where Π_H is the L^2-projection operator onto the space of constants on $T\in\mathcal{T}_H$; see [BF91, Sect. III.3.4]. Inequality (2.12) follows immediately.

The proof of (2.13) uses Green's formula, the norm equivalence (2.10), and a partition of unity very similar to the one given in [DSW94] for the simplicial case. Consider a face $F\subset\partial T$, and note that it is decomposed into non-overlapping faces of the fine triangulation; see Fig. 2.11.

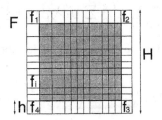

Fig. 2.11. Decomposition of F

Number these faces so that f_i, $1 \leq i \leq n_F$, have at least one vertex on an edge of F; see Fig. 2.11, and let f_1, f_2, f_3, f_4 be the faces that contain a corner point of F. Let $t_i \subset T$, be the associated elements. We remark that, by assumption, the triangulation restricted to the face is quasi-uniform, and thus $n_F \leq C(H/h)$. Let ϑ_F be a continuous, piecewise trilinear function defined on T, which vanishes on $\partial T \setminus F$ and is equal to one at all interior mesh points of F. It equals one in the grey shadowed region in Fig. 2.11. The extension of ϑ_F to the interior of T has values between zero and one, and the absolute value of its gradient is bounded by $C/\max(r, h)$, where r denotes the distance to the wirebasket of T. The wirebasket of T is the union of the twelve edges of T. We refer to [DSW94] for an explicit construction of such a function for a simplex; this construction can easily be adapted to the cubic case. The following upper bound for the H^1-semi norm and the L^2-norm can then be established

$$|\vartheta_F|^2_{1;T} \leq CH_T\left(1 + \log\frac{H}{h}\right), \quad \|\vartheta_F\|^2_{0;T} \leq CH^3_T . \tag{2.14}$$

Using (2.10), it is sufficient to bound the absolute value of $\lambda_F(\rho_H \mathbf{v})$ to get an upper bound for the L^2-norm of $\rho_H \mathbf{v}$. Applying the definition of ρ_H and Green's formula, we obtain

$$|F|\,\lambda_F(\rho_H \mathbf{v}) = \int_F (\rho_H \mathbf{v} \cdot \mathbf{n})\,d\sigma = \int_F (\mathbf{v} \cdot \mathbf{n})\,d\sigma$$

$$= \int_{\partial T} \vartheta_F(\mathbf{v} \cdot \mathbf{n})\,d\sigma + \frac{3}{4}\sum_{i=1}^{4}|f_i|\left(\mathbf{v} \cdot \mathbf{n}_{|f_i}\right) + \frac{1}{2}\sum_{i=5}^{n_F}|f_i|\left(\mathbf{v} \cdot \mathbf{n}_{|f_i}\right)$$

$$= \int_T (\vartheta_F\,\mathrm{div}\,\mathbf{v} + \mathbf{grad}\,\vartheta_F \cdot \mathbf{v})\,dx + \sum_{i=1}^{n_F}\beta_i\,|f_i|\,\lambda_{f_i}(\mathbf{v}) ,$$

where $\beta_i = 3/4$ for $1 \leq i \leq 4$ and $\beta_i = 1/2$ for $5 \leq i \leq n_F$. Thanks to (2.10), the L^2-norm of $\rho_H \mathbf{v}$ can be bounded by

$$H_T\|\rho_H \mathbf{v}\|^2_{0;T} \leq C \sum_{F \subset \partial T}\left(\int_T (\vartheta_F \mathrm{div}\,\mathbf{v} + \mathbf{grad}\,\vartheta_F \cdot \mathbf{v})\,dx\right)^2$$

$$+ C \sum_{F \subset \partial T}\left(\tfrac{3}{4}\sum_{i=1}^{4}|f_i|\lambda_{f_i}(\mathbf{v}) + \tfrac{1}{2}\sum_{i=5}^{n_F}|f_i|\lambda_{f_i}(\mathbf{v})\right)^2 .$$

By means of (2.14), we find an upper bound for the first sum on the right hand side

$$C\left(H^3_T\|\mathrm{div}\,\mathbf{v}\|^2_{0;T} + H_T\left(1 + \log\frac{H}{h}\right)\|\mathbf{v}\|^2_{0;T}\right) . \tag{2.15}$$

Applying (2.10) again and keeping in mind that n_F is bounded by CH/h, we find an upper bound of the second term

$$C \sum_{F \subset \partial T} n_F \sum_{i=1}^{n_F} |f_i|^2 (\lambda_{f_i}(\mathbf{v}))^2 \leq C H_T \|\mathbf{v}\|_{0;T}^2 . \tag{2.16}$$

The upper bound (2.15) and inequality (2.16) finally give

$$\|\rho_H \mathbf{v}\|_{0;T}^2 \leq C H_T^2 \|\operatorname{div} \mathbf{v}\|_{0;T}^2 + C \left(1 + \log \frac{H}{h}\right) \|\mathbf{v}\|_{0;T}^2 . \qquad \square$$

We can obtain a similar estimate for the energy norm on each substructure

$$\int_T B (\rho_H \mathbf{v}) \cdot (\rho_H \mathbf{v}) \, dx \leq C \frac{\gamma_T}{\beta_T} \left(1 + \log \frac{H}{h}\right) \int_T B \, \mathbf{v} \cdot \mathbf{v} \, dx$$
$$+ C \frac{H_T^2 \gamma_T}{a_T} \int_T a \operatorname{div} \mathbf{v} \, \operatorname{div} \mathbf{v} \, dx .$$

Here, a_T is the minimum of $a(x)$ on T, and β_T and γ_T satisfy

$$\beta_T \, \eta^T \eta \leq \eta^T B(x) \eta \leq \gamma_T \, \eta^T \eta, \quad \eta \in \mathbb{R}^3, \, x \in T .$$

Thus, the constant in the corresponding global estimate depends on the ratio of the coefficients B and a on individual substructures, but not on the jumps of the coefficients between the substructures.

Remark 2.5. *The interpolation operator ρ_H is logarithmically stable in the $\|\cdot\|_{\operatorname{div}}$-norm. This result holds for the 3D case as well as in the 2D case; see [TWW00]. In contrast, the nodal interpolant onto continuous finite element spaces has a norm which grows as $(H/h)^{1/2}$ in the 3D case; see [DSW94]. This is an essential difference between the face based Raviart–Thomas finite elements and the standard vertex based Lagrangian finite elements.*

2.2.2.2 An Extension Operator onto $\mathbf{V}_F$. The second basic tool in the analysis of our iterative substructuring method is an extension operator from $W_h(F)$ onto V_F. Here, $W_h(F)$ is the finite element space of piecewise constants on $f \subset F$. The idea is analogous to a result for the standard finite element case, in which the H^1-norm of a discrete harmonic function is bounded by the $H^{1/2}$-norm of its trace on the boundary. The following lemma shows that the extension operator is uniquely defined and stable.

Lemma 2.6. *There exists a unique extension operator $\mathcal{H}_F : W_h(F) \longrightarrow V_F$ satisfying*

$$(\mathcal{H}_F \mu) \cdot \mathbf{n}_{|F} = \mu, \quad \mu \in W_h(F) .$$

Furthermore if $\int_F \mu \, d\sigma = 0$, we have the following stability estimate

$$\|\mathcal{H}_F \mu\|_{0;T}^2 + \|\operatorname{div} \mathcal{H}_F \mu\|_{0;T}^2 \leq C \|\mu\|_{-\frac{1}{2};\partial T}^2, \quad F \subset \partial T ,$$

where μ is extended by zero on $\partial T \setminus F$.

Proof. The degrees of freedom of V_F are exactly the normal components on $f \subset F$, which are constant on each face f. Thus the trace space of normal components of V_F is exactly $W_h(F)$. Therefore, the existence and uniqueness of such an extension operator follows from the definition of V_F.

The proof of the stability estimate is based on a lemma established in [WTW00], where a stable extension to a divergence free vector field is constructed for $\mu \in W_h(F)$ with $\int_F \mu \, d\sigma = 0$. Here, we sketch the ideas of this proof and refer to [WTW00] for the details. The starting point is a Neumann boundary value problem for the Laplace operator on T, where $F \subset \partial T$. We extend μ by zero on $\partial T \setminus F$. The extension is still denoted by μ. Then, the Neumann boundary value problem

$$-\Delta u = 0, \quad \text{in } T, \quad \frac{\partial u}{\partial n} = \mu, \quad \text{on } \partial T \qquad (2.17)$$

has a solution. Moreover, we obtain a unique solution of (2.17) by imposing the additional condition $\int_T u \, dx = 0$. An elementary regularity result gives an upper bound of the H^1-norm of the solution in terms of the boundary condition, i.e., $|u|_{1;T} \leq C\|\mu\|_{-1/2;\partial T}$. The resulting vector field $\mathbf{grad}\, u$ is in $H(\mathrm{div}\,; T)$. In a second step, we apply the natural interpolation operator onto the Raviart–Thomas vector fields, and denote the result by $\mathbf{v}_\mu$. The stability, $\|\mathbf{v}_\mu\|_{\mathbf{div};T} \leq C|u|_{1;T}$, is a consequence of an approximation property and an inverse estimate. By construction $\mathbf{v}_\mu$ is divergence free, and it has the same normal trace as $\mathcal{H}_F\mu$ on ∂T. Thus, $\mathcal{H}_F\mu$, restricted to T, can be written as $\mathcal{H}_F\mu = \mathbf{v}_\mu + \mathbf{v}_T$, where $\mathbf{v}_T \in V_T$. Keeping, the definition of V_F in mind, we find that

$$a_T(\mathcal{H}_F\mu, \mathcal{H}_F\mu) = a_T(\mathcal{H}_F\mu, \mathcal{H}_F\mu - \mathbf{v}_T) \leq a_T(\mathcal{H}_F\mu, \mathcal{H}_F\mu)^{\frac{1}{2}} a_T(\mathbf{v}_\mu, \mathbf{v}_\mu)^{\frac{1}{2}} ,$$

where $a_T(\cdot, \cdot)$ is the restriction of $a(\cdot, \cdot)$ on T. $\quad\square$

As already mentioned, an important step in the proof of a sharp bound for C_0 is the choice of $\mathbf{v}_H$. Defining $\mathbf{v}_H := \rho_H \mathbf{v}$, we find

$$\mathbf{v}_F = \mathcal{H}_F((\mathbf{v} - \rho_H \mathbf{v}) \cdot \mathbf{n}_{|_F})$$

and thus, by means of Lemma 2.6, we get an upper bound for the norm of $\mathbf{v}_F$ in terms of the normal components of $\mathbf{v} - \rho_H \mathbf{v}$ restricted to F

$$a_T(\mathbf{v}_F, \mathbf{v}_F) \leq C \, \|\mu\|^2_{-\frac{1}{2};\partial T} ,$$

where $\mu_{|_F} = ((\mathbf{v} - \rho_H \mathbf{v}) \cdot \mathbf{n})_{|_F}$ and $\mu = 0$ on $\partial T \setminus F$. To apply Lemma 2.1, an estimate of the $H^{-1/2}$-norm of μ in terms of the norm of $\mathbf{v}$ is required. We conclude this paragraph by proving a decomposition lemma for piecewise constant functions on the boundary of a substructure. We recall that this space is exactly the trace space of normal components of Raviart–Thomas finite elements.

The following lemma is established in [WTW00]. Here, we briefly discuss the main ideas and omit some technical details. In the following each element μ_F in $W_h(F)$ is extended by zero onto $\partial T \setminus F$ and is still denoted by μ_F. The subspace of $W_h(F)$ with zero mean value on F is called $W_{0;h}(F) := \{\mu \in W_h(F), \int_F \mu \, d\sigma = 0\}$, and the six dimensional space on ∂T containing all functions which are constant on each face F is denoted by $W_H(\partial T)$.

Lemma 2.7. *Let T be in $\mathcal{T}_H$, let $\mu_F \in W_{0;h}(F)$, $F \subset \partial T$, and let $\mu := \sum_{F \subset \partial T} \mu_F$. Then, there exists a constant C, independent of $\mu_H \in W_H(\partial T)$ and h such that*

$$\|\mu_F\|^2_{-\frac{1}{2};\partial T} \leq C\left(1 + \log \frac{H}{h}\right)\left((1 + \log \frac{H}{h})\|\mu + \mu_H\|^2_{-\frac{1}{2};\partial T} + \|\mu\|^2_{-\frac{1}{2};\partial T}\right) . \tag{2.18}$$

Proof. By definition, the mean value on ∂T of μ_F is zero. Applying a Poincaré–Friedrich's type inequality, we obtain

$$\|\mu_F\|_{-\frac{1}{2};\partial T} = \sup_{\phi \in H^{\frac{1}{2}}(\partial T)} \frac{\langle \mu_F, \phi \rangle}{\|\phi\|_{\frac{1}{2};\partial T}} = \sup_{\phi \in H^{\frac{1}{2}}(\partial T)} \frac{\langle \mu_F, \phi - c_\phi \rangle}{\|\phi\|_{\frac{1}{2};\partial T}}$$

$$\leq C \sup_{\substack{\phi \in H^{\frac{1}{2}}(\partial T) \\ \phi \neq \text{const.}}} \frac{\langle \mu_F, \phi \rangle}{|\phi|_{-\frac{1}{2};\partial T}} ,$$

where c_ϕ is the mean value of ϕ on ∂T. The essential idea is now to replace the supremum over $H^{1/2}(\partial T)$ by the supremum over a suitable discrete space. We introduce a space of bubbles on ∂T by

$$B_h(\partial T) := \{\phi \in L^2(\partial T)| \phi_{|_f} = \alpha_f \lambda_1 \lambda_2 \lambda_3 \lambda_4, \ f \subset \partial T, \alpha_f \in \mathbb{R}\} ,$$

where λ_i, $1 \leq i \leq 4$, are the barycentric coordinate functions of the face f. We also use the space of conforming bilinear finite elements on ∂T

$$S_h(\partial T) := \{\phi \in C(\partial T)| \phi_{|_f} \in Q(f), \ f \subset \partial T\} ,$$

where $Q(f)$ is the space of bilinear functions on f. Let $P : H^{1/2}(\partial T) \longrightarrow B_h(\partial T) + S_h(\partial T)$ be a uniform $H^{1/2}$-stable operator which satisfies

$$\int_F \phi \, \mu_F \, d\sigma = \int_F P\phi \, \mu_F \, d\sigma, \quad \mu_F \in W_h(F), \ F \subset \partial T . \tag{2.19}$$

For the existence of such an operator, it is sufficient to construct one example. We set $P\phi := \phi_S + \phi_B$, where $\phi_S := P_S \, \phi$ is given by a standard locally defined $H^{1/2}$-stable quasi-projection operator P_S onto $S_h(\partial T)$; see, e.g., [SZ90], and $\phi_B \in B_h(\partial T)$ is defined by

$$\int_f \phi_B \, d\sigma = \int_f (\phi - \phi_S) \, d\sigma, \quad f \subset \partial T .$$

Then, P satisfies (2.19) by construction. Furthermore, the $H^{1/2}$-semi norm of $P\phi$ can be bounded by $|P\phi|_{1/2;\partial T} \leq |\phi_S|_{1/2;\partial T} + |\phi_B|_{1/2;\partial T}$. If the quasi-projection operator P_S satisfies an approximation property, we find for the second term

$$|\phi_B|_{\frac{1}{2};\partial T}^2 \leq \frac{C}{h}\|\phi_B\|_{0;\partial T}^2 \leq \frac{C}{h}\|\phi - \phi_S\|_{0;\partial T}^2 \leq C|\phi|_{\frac{1}{2};\partial T}^2$$

which yields the $H^{1/2}$-stability of P. Moreover, the spaces $S_h(\partial T)$ and $B_h(\partial T)$ satisfy an orthogonality relation, and we find $(\nabla\psi_S, \nabla\Psi_B)_{0;\partial T} = 0$, $\psi_S \in S_h(\partial T)$, $\psi_B \in B_h(\partial T)$. Then, an interpolation argument and an inverse estimates yield

$$|\psi_B|_{\frac{1}{2};\partial T}^2 \leq Ch|\psi_B|_{1;\partial T}^2 \leq Ch|\psi_B + \psi_S|_{1;\partial T}^2 \leq C|\psi_B + \psi_S|_{\frac{1}{2};\partial T}^2 \ .$$

These preliminary considerations guarantee the following norm equivalence

$$c\,|\psi_S + \psi_B|_{\frac{1}{2};\partial T} \leq |\psi_S|_{\frac{1}{2};\partial T} + |\psi_B|_{\frac{1}{2};\partial T} \leq C\,|\psi_S + \psi_B|_{\frac{1}{2};\partial T} \ . \qquad (2.20)$$

Now, the supremum over $H^{1/2}(\partial T)$ can be modified. In particular, we can replace the continuous space $H^{1/2}(\partial T)$ by the discrete spaces $S_h(\partial T)$ and $B_h(\partial T)$, and we find the following upper bound

$$\|\mu_F\|_{-\frac{1}{2};\partial T} \leq C \sup_{\substack{\phi \in H^{\frac{1}{2}}(\partial T) \\ \phi \neq \text{const.}}} \frac{\langle \mu_F, P\phi \rangle}{|\phi|_{\frac{1}{2};\partial T}} \leq C \sup_{\substack{\phi \in S_h(\partial T) + B_h(\partial T) \\ \phi \neq \text{const.}}} \frac{\langle \mu_F, \phi \rangle}{|\phi|_{\frac{1}{2};\partial T}}$$

$$\leq C \left(\sup_{\substack{\phi \in B_h(\partial T) \\ \phi \neq 0}} \frac{\langle \mu_F, \phi \rangle}{|\phi|_{\frac{1}{2};\partial T}} + \sup_{\substack{\phi \in S_h(\partial T) \\ \phi \neq \text{const.}}} \frac{\langle \mu_F, \phi \rangle}{|\phi|_{\frac{1}{2};\partial T}} \right) \ . \qquad (2.21)$$

The first inequality is a consequence of (2.19), whereas the second follows from the stability in the $H^{1/2}$-semi norm of P. Finally, the last inequality follows by using (2.20).

We next consider the two terms on the right hand side of (2.21) separately, and start with the first term. A local inverse estimate and an interpolation argument shows that the restriction ϕ_F of $\phi \in B_h(\partial T)$ to F is $H^{1/2}$-stable,

$$|\phi_F|_{\frac{1}{2};\partial T} \leq C|\phi|_{\frac{1}{2};\partial T} \ ,$$

where $\phi_F := \phi$ on F and $\phi_F := 0$ on $\partial T \setminus F$. Using this bound, we find that for $\phi \in B_h(\partial T)$

$$\frac{|\langle \mu_F, \phi \rangle|}{|\phi|_{\frac{1}{2};\partial T}} = \frac{|\langle \mu, \phi_F \rangle|}{|\phi|_{\frac{1}{2};\partial T}} \leq \frac{\|\mu\|_{-\frac{1}{2};\partial T}\,\|\phi_F\|_{\frac{1}{2};\partial T}}{|\phi|_{\frac{1}{2};\partial T}}$$

$$\leq C\frac{\|\mu\|_{-\frac{1}{2};\partial T}\,|\phi_F|_{\frac{1}{2};\partial T}}{|\phi|_{\frac{1}{2};\partial T}} \leq C\|\mu\|_{-\frac{1}{2};\partial T} \ . \qquad (2.22)$$

Unfortunately, the second term in (2.21) cannot be bounded as easily. We start by defining for each $\phi \in S_h(\partial T)$ a weighted average c_ϕ

$$c_\phi \int_F \vartheta_F \, d\sigma := \int_F I_h(\vartheta_F \phi) \, d\sigma \ .$$

Here, ϑ_F is given in the proof of Lemma 2.4, and I_h is the nodal interpolation operator onto $S_h(\partial T)$. Then, the supremum in the second term on the right hand side in (2.21) can be replaced by

$$\sup_{\substack{\phi \in S_h(\partial T) \\ \phi \neq \text{const.}}} \frac{\langle \mu_F, \phi \rangle}{|\phi|_{\frac{1}{2};\partial T}} = \sup_{\substack{\phi \in S_h(\partial T) \\ \phi \neq \text{const.}}} \frac{\langle \mu_F, \phi - c_\phi \rangle}{|\phi - c_\phi|_{\frac{1}{2};\partial T}} \leq C \sup_{\substack{\phi \in S_h(\partial T) \\ \phi \neq 0, \, c_\phi = 0}} \frac{\langle \mu_F, \phi \rangle}{\|\phi\|_{\frac{1}{2};\partial T}} \ , \quad (2.23)$$

i.e., we need only consider functions ϕ which have a zero weighted mean value on F. In the last step, we have used the following norm equivalence

$$c(|\phi|_{\frac{1}{2};\partial T}^2 + H c_\phi^2) \leq \|\phi\|_{\frac{1}{2};\partial T}^2 \leq C(|\phi|_{\frac{1}{2};\partial T}^2 + H c_\phi^2) \ .$$

This is a Poincaré–Friedrich's type inequality, and can be easily proved by contradiction.

We can now apply techniques well established for the standard Lagrangian finite elements in 3D; see [DSW94]. In general, $\phi \in S_h(\partial T)$ cannot be decomposed in a uniformly stable way into face and edge contributions. Therefore, we start by decomposing ϕ into a sum of contributions ϕ_F supported on individual faces $F \subset \partial T$, and ϕ_w supported in a neighborhood of the wirebasket which is one element wide; see Fig. 2.12.

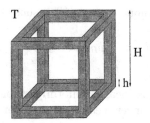

Fig. 2.12. Neighborhood of the wirebasket

By means of this decomposition, we can rewrite ϕ as

$$\phi = \sum_{F \subset \partial T} \phi_F + \phi_w \ . \quad (2.24)$$

The $H^{1/2}$-semi norm on ∂T of ϕ_w can be bounded by means of an inverse estimate in term of its one dimensional L^2-norm over the wirebasket W

$$|\phi_w|_{\frac{1}{2};\partial T}^2 \leq C \frac{1}{h} \|\phi_w\|_{0;\partial T}^2 \leq C \|\phi_w\|_{0;W}^2 \ .$$

Upper bounds for $\|\phi_w\|_{0;W}^2$ and $\|\phi_F\|_{1/2;\partial T}^2$ are established for the simplicial case in [DSW94]. Following [DSW94, Lemma 4.3 and Lemma 4.5], and observing that the H^1-norm on T of a discrete harmonic function can be bounded by the $H^{1/2}$-norm of its trace on ∂T, it can be shown that

$$\|\phi_w\|_{0;W}^2 \leq C\left(1 + \log \frac{H}{h}\right) \|\phi\|_{\frac{1}{2};\partial T}^2 , \qquad (2.25)$$

$$\|\phi_F\|_{\frac{1}{2};\partial T}^2 \leq C\left(1 + \log \frac{H}{h}\right)^2 \|\phi\|_{\frac{1}{2};\partial T}^2 . \qquad (2.26)$$

The proofs in [DSW94] are for the simplicial case, but they can be carried out in exactly the same way for the hexahedral case and details are therefore omitted. We find, by using the splitting (2.24), that

$$\langle \mu_F, \phi \rangle = \sum_{\hat{F} \subset \partial T} \langle \mu_F, \phi_{\hat{F}} \rangle + \langle \mu_F, \phi_w \rangle = \langle \mu, \phi_F \rangle + \langle \mu_F, \phi_w \rangle . \qquad (2.27)$$

Since $I_h(\vartheta_F \phi) = \phi_F = I_h(\vartheta_F \phi_F)$ on ∂T, and since we can always assume that $c_\phi = 0$, we obtain

$$\langle \mu_H, \phi_F \rangle = 0, \quad \mu_H \in W_H(\partial T) ,$$

and $c_{\phi_F} = 0$. The first term on the right side of (2.27) can be bounded by means of (2.26):

$$|\langle \mu, \phi_F \rangle| = |\langle \mu + \mu_H, \phi_F \rangle| \leq C\left(1 + \log \frac{H}{h}\right)\|\phi\|_{\frac{1}{2};\partial T} \|\mu + \mu_H\|_{-\frac{1}{2};\partial T} . \quad (2.28)$$

For each ϕ_w, there is a unique $\phi_B \in B_h(\partial T)$ such that $\phi_B = 0$ on $\partial T \setminus F$ and

$$\int_f \phi_w \, d\sigma = \int_f \phi_B \, d\sigma, \quad f \in \mathcal{F}_h, \ f \subset F .$$

Moreover, this mapping is locally uniformly stable in the L^2-norm on ∂T; $\|\phi_B\|_{0;f} \leq C\|\phi_w\|_{0;f}$. An inverse estimate together with the definition of ϕ_B easily yield

$$\|\phi_B\|_{\frac{1}{2};\partial T}^2 \leq C\frac{1}{h}\|\phi_B\|_{0;\partial T}^2 \leq C\frac{1}{h}\|\phi_w\|_{0;\partial T}^2 \leq C \|\phi_w\|_{0;W}^2 .$$

By means of this bound and (2.25), we finally obtain

$$|\langle \mu_F, \phi_w \rangle| = |\langle \mu_F, \phi_B \rangle| = |\langle \mu, \phi_B \rangle| \leq C\|\mu\|_{-\frac{1}{2};\partial T}\|\phi_w\|_{0;W}$$

$$\leq C\left(1 + \log \frac{H}{h}\right)^{\frac{1}{2}}\|\mu\|_{-\frac{1}{2};\partial T}\|\phi\|_{\frac{1}{2};\partial T} . \qquad (2.29)$$

The proof is completed by combining (2.21), (2.22), (2.23), (2.24), (2.28), and (2.29).   □

2.2.2.3 Quasi-optimal Bounds. We are now able to formulate the central result of this section. In the following theorem, we specify a suitable decomposition of $\mathbf{v} \in V_h$ and give an upper bound of the constant C_0 in Lemma 2.1. This result was originally given in [WTW00].

Theorem 2.8. *For each $\mathbf{v} \in V_h$, there exists a decomposition*

$$\mathbf{v} = \mathbf{v}_H + \sum_{T \in \mathcal{T}_H} \mathbf{v}_T + \sum_{F \in \mathcal{F}_H} \mathbf{v}_F \ ,$$

corresponding to (2.11), such that

$$a(\mathbf{v}_H, \mathbf{v}_H) + \sum_{T \in \mathcal{T}_H} a(\mathbf{v}_T, \mathbf{v}_T) + \sum_{F \in \mathcal{F}_H} a(\mathbf{v}_F, \mathbf{v}_F) \leq C \left(1 + \log \frac{H}{h}\right)^2 a(\mathbf{v}, \mathbf{v}) \ ,$$

with a constant C, independent of h, H, and $\mathbf{v}$.

Proof. The choice $\mathbf{v}_H := \rho_H \mathbf{v}$ and Lemma 2.4 yield

$$a_T(\mathbf{v}_H, \mathbf{v}_H) \leq C \left(1 + \log \frac{H}{h}\right) a_T(\mathbf{v}, \mathbf{v}), \quad \mathbf{v} \in V_h \ ,$$

where $a_T(\cdot, \cdot)$ is the restriction of $a(\cdot, \cdot)$ on one substructure T. Then, $\mathbf{v}_F$ is uniquely given as the discrete harmonic extension of $\mu_F := (\mathbf{v} - \mathbf{v}_H) \cdot \mathbf{n}|_F$. The minimization property of $\mathcal{H}_F$ gives

$$a_T(\mathbf{v}_F, \mathbf{v}_F) \leq C \|\mu_F\|^2_{-\frac{1}{2}; \partial T}, \quad F \subset \partial T \ ;$$

see Lemma 2.6. In a final step, we apply Lemma 2.7 and bound $\|\mu_F\|_{-1/2; \partial T}$ in terms of $a_T(\mathbf{v}, \mathbf{v})^{1/2}$. Setting $\mu_H := \mathbf{v}_H \cdot \mathbf{n}$, we obtain

$$a_T(\mathbf{v}_F, \mathbf{v}_F) \leq C\left(1 + \log \frac{H}{h}\right)\left(\|(\mathbf{v} - \mathbf{v}_H) \cdot \mathbf{n}\|^2_{-\frac{1}{2}; \partial T} + \left(1 + \log \frac{H}{h}\right)\|\mathbf{v} \cdot \mathbf{n}\|^2_{-\frac{1}{2}; \partial T}\right) \ .$$

Lemma 2.3, the triangle inequality, and Lemma 2.4 yield

$$\|(\mathbf{v} - \mathbf{v}_H) \cdot \mathbf{n}\|^2_{-\frac{1}{2}; \partial T} \leq C\left(1 + \log \frac{H}{h}\right) a_T(\mathbf{v}, \mathbf{v}) \ ,$$

and thus by applying Lemma 2.3 to $\|\mathbf{v} \cdot \mathbf{n}\|_{-1/2; \partial T}$, we find

$$a_T(\mathbf{v}_F, \mathbf{v}_F) \leq C\left(1 + \log \frac{H}{h}\right)^2 a_T(\mathbf{v}, \mathbf{v}) \ .$$

An upper bound for $a_T(\mathbf{v}_T, \mathbf{v}_T)$ is now an easy consequence of the triangle inequality. Finally, the global upper estimate is obtained by summing the local ones over the elements of the macro-triangulation.   □

Remark 2.9. *The constant C in Theorem 2.8 depends on the coefficients but not on the meshsize. Since all estimates are done locally on individual substructures, it can be shown by a more careful analysis that C is independent on the jumps of the coefficients. However, it depends linearly on*

$$\max_{T \in \mathcal{T}_H} \frac{\gamma_T}{\beta_T} \left(1 + \frac{\gamma_T H^2}{a_T}\right) .$$

Using exact orthogonal projections onto the subspaces, the additive Schwarz operator T_{add}, defined by the decomposition (2.11), has a condition number which is bounded logarithmically in terms of the ratio H/h

$$\kappa(T_{\mathrm{add}}) \leq C \left(1 + \log \frac{H}{h}\right)^2 ,$$

where the constant C does not depend on the jumps of the coefficients. This is an easy consequence of Lemma 2.1, Theorem 2.8, and a coloring argument.

2.2.3 A Hierarchical Basis Method

In this subsection, we introduce a hierarchical basis method for Raviart–Thomas finite elements in 3D. We obtain the same qualitative upper bound for the condition number of the additive Schwarz method as for standard Lagrangian finite elements in 2D; see, e.g., [Ban96, Yse86, Yse93]. We recall that this result does not hold in 3D for the Lagrangian finite elements.

The starting point for the definition of our hierarchical basis method is a nested sequence of adaptively generated simplicial triangulations, $\mathcal{T}_0, \cdots \mathcal{T}_j$. We can use some standard refinement rules guaranteeing that the triangulations form a family of shape regular and locally quasi-uniform triangulations; see, e.g., [Bey95]. The sets of interior faces are denoted by $\mathcal{F}_l$, $0 \leq l \leq j$, and the associated Raviart–Thomas finite element spaces are called V_l, $0 \leq l \leq j$. They form a sequence of nested spaces satisfying $V_0 \subset V_1 \subset \ldots \subset V_j \subset H_0(\mathrm{div}\,;\Omega)$. The local Raviart–Thomas spaces with four degrees of freedom per element are defined by (2.9), and the global space has one degree of freedom per interior face.

To define our Schwarz method, we consider two different types of decompositions – a horizontal and a vertical. The horizontal decomposition is based on the hierarchy of finite element spaces whereas the vertical one reflects a Helmholtz-type decomposition of the vector fields. This type of splitting has already been used in [HW97, Woh95] for the 2D case. We also note that a hierarchical basis preconditioner for the saddle point problem arising from the mixed finite element discretization of an elliptic second order operator has been introduced in [Woh95]. Similar decomposition techniques are used in [Hip96, Hip97] to construct an optimal multigrid method for Raviart–Thomas finite elements in 3D based on a uniformly refined sequence of triangulations.

Efficient preconditioners and multigrid techniques are discussed and analyzed in [AFW97, AFW98, AFW00].

The natural interpolation operator $\rho_l : V_j \longrightarrow V_l$, $0 \le l \le j$, will play an essential role in the analysis of our hierarchical basis method. For the convenience of the reader, we recall the definition already given in the previous subsection

$$\lambda_F(\rho_l(\mathbf{v})) := \frac{1}{|F|} \int_F \mathbf{v} \cdot \mathbf{n} \, d\sigma, \quad F \in \mathcal{F}_l \ ,$$

where $\lambda_F(\cdot)$ are the degrees of freedom as given in Subsect. 2.2.1. We remark that $\rho_j = \mathrm{Id}$, and we define $\rho_{-1} := 0$.

2.2.3.1 Horizontal Decomposition. We define a family of subspaces $\widetilde{V}_l$, $0 \le l \le j$, of V_j by

$$\widetilde{V}_l := (\rho_l - \rho_{l-1})V_j \ , \quad 0 \le l \le j.$$

Then, obviously $\widetilde{V}_0 = V_0$ and $\widetilde{V}_l \subset V_l$ and $\widetilde{V}_l \cap \widetilde{V}_k = 0$, $0 \le l \ne k \le j$. Furthermore, each element $\mathbf{v} \in V_j$ can be written as

$$\mathbf{v} = \sum_{l=0}^{j}(\rho_l - \rho_{l-1})\mathbf{v} =: \sum_{l=0}^{j} \mathbf{v}_l \ ,$$

and thus

$$V_j = \sum_{l=0}^{j} \widetilde{V}_l \tag{2.30}$$

is a direct sum. This decomposition already defines an additive Schwarz method if the exact projections are used. In a first step, we show that the constant C_0 in Lemma 2.1 can be bounded quadratically in j and in a second step that the spectral radius of $\mathcal{E}$ in Lemma 2.1 is bounded by a constant independent of j. The following lemma gives an upper bound for C_0 and can be obtained easily from Lemma 2.4.

Lemma 2.10. *There exists a constant independent of j such that*

$$a(\mathbf{v}_l, \mathbf{v}_l) \le C(1 + l) \, a(\mathbf{v}, \mathbf{v}), \quad 0 \le l \le j \ .$$

Proof. The proof is based on Lemma 2.4. Using the triangle inequality, we find

$$a(\mathbf{v}_l, \mathbf{v}_l) \le 2(a(\rho_l \mathbf{v}, \rho_l \mathbf{v}) + a(\rho_{l-1}\mathbf{v}, \rho_{l-1}\mathbf{v})) \ .$$

In the case of the iterative substructuring method, we assumed that the fine triangulation is obtained by quasi-uniform refinement of a coarse triangulation. Thus before we can apply Lemma 2.4, we have to make some modifications. We introduce a fictitious sequence of nested triangulations $\widehat{\mathcal{T}}_0, \widehat{\mathcal{T}}_1, \cdots \widehat{\mathcal{T}}_j$

such that each element in $\mathcal{T}_l$, $0 \le l \le j$, can be written as a union of elements in $\widehat{\mathcal{T}}_l$. An explicit construction for the 2D case is given in [Woh95]. Figure 2.13 illustrates the construction of $\widehat{\mathcal{T}}_l$ in 2D. The first row in Fig. 2.13 shows the sequence $\mathcal{T}_l$ and the second row the corresponding sequence $\widehat{\mathcal{T}}_l$.

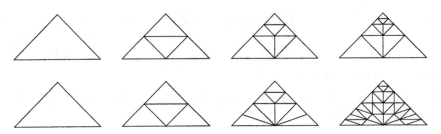

Fig. 2.13. Construction of the sequence $\widehat{\mathcal{T}}_l$ in 2D

To obtain $\widehat{\mathcal{T}}_l$, we refine each element, $T \in \mathcal{T}_0$, l times quasi-uniformly. Here, regular refinement into eight subelements as well as refinement based on bisection can be applied; see, e.g., [Bae91, Bey95]. We insist that a refinement step based on bisection is carried out at most one time during the l steps, and thus the shape regularity of the triangulations is preserved. Then, each element in $\widehat{\mathcal{T}}_l$ which is a subset of $T \in \mathcal{T}_0$ has a diameter bounded from above and below by constants times $2^{-l}H_T$, where H_T is the diameter of T, and $\log(H/h)$ is bounded from above by Cl. We remark that each element in $\mathcal{T}_l$, which is not at the same time an element in $\widehat{\mathcal{T}}_l$, is also an element in $\mathcal{T}_j$.

We associate with $\widehat{\mathcal{T}}_l$ the Raviart–Thomas finite element spaces $\widehat{V}_l$, and find by construction that $V_l \subset \widehat{V}_l$ and moreover that $\hat{\rho}_l\mathbf{v} = \rho_l\mathbf{v}$ for $\mathbf{v} \in V_j$. Here, $\hat{\rho}_l$ is the natural interpolant onto $\widehat{V}_l$. Observing that $\widehat{\mathcal{T}}_l$ satisfies the assumption of Subsect. 2.2.2, we can apply Lemma 2.4 and obtain

$$a(\mathbf{v}_l, \mathbf{v}_l) \le 2(a(\hat{\rho}_l\mathbf{v}, \hat{\rho}_l\mathbf{v}) + a(\hat{\rho}_{l-1}\mathbf{v}, \hat{\rho}_{l-1}\mathbf{v})) \le C(1+l)\, a(\mathbf{v}, \mathbf{v})\ .$$

$\square$

Applying Lemma 2.10 and summing over the refinement levels, we obtain an upper bound for C_0

$$\sum_{l=0}^{j} a(\mathbf{v}_l, \mathbf{v}_l) \le C(1+j^2)\, a(\mathbf{v}, \mathbf{v})\ . \tag{2.31}$$

To prove an upper bound for the condition number, we have to consider $\mathcal{E}$ in more detail since establishing a sharp bound for $\rho(\mathcal{E})$ is not as easy as in case of overlapping or iterative substructuring methods. To get an upper bound for $\rho(\mathcal{E})$, which is independent of the number of refinement levels, we have to show a suitable strengthened Cauchy–Schwarz inequality.

Lemma 2.11. *The following Cauchy–Schwarz inequality holds*

$$(\mathbf{v}_l, \mathbf{v}_k)_{\mathrm{div};T} \leq C 2^{-|l-k|/2} \|\mathbf{v}_l\|_{\mathrm{div};T} \|\mathbf{v}_k\|_{\mathrm{div};T}, \quad \mathbf{v}_l \in \tilde{V}_l, \mathbf{v}_k \in \tilde{V}_k .$$

Proof. Without loss of generality, we assume that $l > k$. We consider one element $T \in \mathcal{T}_k$ at a time, and observe that $\mathbf{v}_{k|_T} \in \mathcal{RT}(T)$. In a first step, we show that the divergence of $\mathbf{v}_k$ and of $\mathbf{v}_l$ are orthogonal with respect to the L^2-scalar product

$$\int_T \mathrm{div}\,\mathbf{v}_l \,\mathrm{div}\,\mathbf{v}_k \, dx = \int_T \mathrm{div}\,(\rho_l \mathbf{v} - \rho_{l-1} \mathbf{v}) \,\mathrm{div}\,\mathbf{v}_k \, dx$$
$$= \int_T (\Pi_l(\mathrm{div}\,\mathbf{v}) - \Pi_{l-1}(\mathrm{div}\,\mathbf{v}))\,\mathrm{div}\,\mathbf{v}_k \, dx = 0 .$$

Here, Π_l and Π_{l-1} are the L^2-projections onto the spaces of piecewise constants associated with the triangulations $\mathcal{T}_l$ and $\mathcal{T}_{l-1}$. To get an upper bound for $(\mathbf{v}_l, \mathbf{v}_k)_{\mathrm{div};T}$, we have to consider in a second step the contributions of the L^2-term. Observing that $\mathbf{v}_{k|_T}$ can be written as a gradient of a quadratic function with mean value zero, $\mathbf{v}_{k|_T} = \mathbf{grad}\,\phi$, $\phi \in P_2(T)$, we find by applying Green's formula

$$\int_T \mathbf{v}_l \cdot \mathbf{v}_k \, dx = -\int_T \mathrm{div}\,\mathbf{v}_l\,\phi\, dx + \int_{\partial T} \phi\, \mathbf{v}_l \cdot \mathbf{n}\, d\sigma .$$

By the definitions of $\mathbf{v}_l$ and $\widehat{\mathcal{T}}_{l-1}$, we get the following orthogonality relations

$$\int_T \mathrm{div}\,\mathbf{v}_l\,\widehat{\Pi}_{l-1}\phi\, dx = 0 ,$$

where $\widehat{\Pi}_{l-1}$ is the L^2-projection onto the space of piecewise constants associated with the triangulation $\widehat{\mathcal{T}}_{l-1}$, and

$$\int_{\partial T} \widehat{\Pi}_{l-1}\phi\, \mathbf{v}_l \cdot \mathbf{n}\, d\sigma = 0 .$$

By means of these orthogonality relations and the discrete norm equivalence (2.10), we obtain the following upper bound

$$\int_T \mathbf{v}_l \cdot \mathbf{v}_k \, dx = -\int_T \mathrm{div}\,\mathbf{v}_l(\phi - \widehat{\Pi}_{l-1}\phi)\, dx + \int_{\partial T} (\phi - \widehat{\Pi}_{l-1}\phi)\mathbf{v}_l \cdot \mathbf{n}\, d\sigma$$
$$\leq C\, 2^{-(l-1-k)}\,(h_T\|\mathrm{div}\,\mathbf{v}_l\|_{0;T}\|\mathbf{v}_k\|_{0;T} + h_T^{\frac{1}{2}}\|\mathbf{v}_l \cdot \mathbf{n}\|_{0;\partial T}\|\mathbf{v}_k\|_{0;T})$$
$$\leq C\, 2^{-(l-k)/2}(\|\mathrm{div}\,\mathbf{v}_l\|_{0;T} + \|\mathbf{v}_l\|_{0;T})\|\mathbf{v}_k\|_{0;T} .$$

$\square$

An upper bound for $\rho(\mathcal{E})$ can be now derived by using the equivalence between the energy norm and the Hilbert space norm and Lemma 2.11. It is bounded by a constant independent of the number of refinement levels

$$\rho(\mathcal{E}) \leq C \ .$$

Combining the estimate for $\rho(\mathcal{E})$ and Lemma 2.10 with Lemma 2.1, we find that the condition number of this additive Schwarz method with exact projections onto the subspaces is bounded quadratically in j

$$\kappa(T_{\text{add}}) \leq C(1 + j^2) \ .$$

The computational cost for the application of the exact projection is in practice too high. Thus, the exact projection onto the subspaces will, very often, be replaced by quasi-projection operators T_l onto $\tilde{V}_l$. We define our quasi-projections T_l in terms of a vertical decomposition of $\tilde{V}_l$, $1 \leq l \leq j$, and show that the condition number is still bounded quadratically in terms of j. In particular, we prove that (2.3) is satisfied with ω independent of j.

2.2.3.2 Vertical Decomposition. We decompose the global space $\tilde{V}_l$ into local low dimensional spaces which are associated with the faces and elements of the triangulation on level $l - 1$. This splitting is motivated by a Helmholtz-type decomposition. In case of a Helmholtz decomposition, the space would be written as a direct sum of a divergence free subspace and its orthogonal complement. Here, we decompose $\tilde{V}_l$ into a sum of divergence free subspaces plus a kind of quasi-orthogonal complement which is called, in the following, a surplus space. We note that the divergence free Raviart–Thomas vector fields can be obtained from Nédélec finite elements; see [Néd82]. The family of Nédélec finite elements form a subspace of the Hilbert space $H(\mathbf{curl}\,; \Omega)$

$$H(\mathbf{curl}\,; \Omega) := \left\{ \mathbf{q} \in (L^2(\Omega))^3 \,|\, \mathbf{curl}\, \mathbf{q} \in (L^2(\Omega))^3 \right\} \ .$$

We work with the Nédélec finite elements of lowest order which are defined by

$$\mathcal{ND}(\Omega; \mathcal{T}_l) := \left\{ \mathbf{q} \in H(\mathbf{curl}\,; \Omega) |\quad \mathbf{q}_{|_T} \in \mathcal{ND}(T), T \in \mathcal{T}_l \right\}, \quad 0 \leq l \leq j \ ,$$

where $\mathcal{ND}(T)$ stands for the local Nédélec space. In the case of a simplicial triangulation, the local space has dimension six

$$\mathcal{ND}(T) := \begin{pmatrix} \alpha_1 + \beta_1 x \\ \alpha_2 + \beta_2 y \\ \alpha_3 + \beta_3 z \end{pmatrix} \ .$$

The degrees of freedom of the global space are given in terms of the tangential components on the edges

$$\lambda_e(\mathbf{q}) := \frac{1}{|e|} \int_e \mathbf{q} \cdot \mathbf{t}\, ds \ ,$$

where e is an edge of the triangulation $\mathcal{T}_l$, and the direction of $\mathbf{t}$ depends on the direction of $\mathbf{n}$. We now introduce, for each face $F \in \mathcal{F}_{l-1}$ which is not an element of $\mathcal{F}_l$, a subspace of at most three degrees of freedom

$$\mathcal{ND}_{l;F} := \{\mathbf{q} \in \mathcal{ND}(\Omega; \mathcal{T}_l)|\ \lambda_e(\mathbf{q}) = 0, e \not\subset F\}\ ;$$

see Fig. 2.14. Moreover, we have $\mathbf{curl}\,\mathcal{ND}(\Omega; \mathcal{T}_l) \subset V_l$.

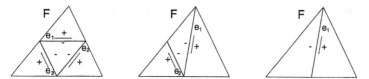

Fig. 2.14. Different refinement techniques on $F \in \mathcal{F}_{l-1}$

We define our divergence free Raviart–Thomas subspaces associated with the faces in terms of the Nédélec spaces $\mathcal{ND}_{l;F}$ by

$$\tilde{V}_{l;F} := \mathbf{curl}\,\mathcal{ND}_{l;F}\ .$$

It can be easily seen that $\tilde{V}_{l;F}$ is a subspace of $\tilde{V}_l$ supported in two elements of $\mathcal{T}_{l-1}$. An element $\mathbf{v}_l \in V_l$ is also in $\tilde{V}_l$ if and only if $\int_F \mathbf{v}_l \cdot \mathbf{n}\, d\sigma = 0$, for $F \in \mathcal{F}_{l-1}$. Let $\mathbf{v} = \mathbf{curl}\,\mathbf{q}$ with $\mathbf{q} \in \mathcal{ND}_{l;F}$, then

$$\int_F \mathbf{v} \cdot \mathbf{n}\, d\sigma = \int_{\partial F} \mathbf{q} \cdot \mathbf{t}\, ds = 0, \quad F \in \mathcal{F}_{l-1}\ .$$

The surplus spaces $\tilde{V}_{l;T}$, $T \in \mathcal{T}_{l-1} \setminus \mathcal{T}_l$, will be defined locally by

$$\tilde{V}_{l;T} := \{\mathbf{v} \in \tilde{V}_l|\ \mathrm{supp}\,\mathbf{v} \subset \overline{T}\}\ .$$

The dimension of $\tilde{V}_{l;T}$ is given by the number of faces $F \in \mathcal{F}_l$ which are in the interior of T; it is bounded by eight. We point out that $\tilde{V}_{l;T}$ can also contain a divergence free element. It has such a one dimensional divergence free subspace, in the case that $T \in \mathcal{T}_{l-1}$ is refined in such a way that it has an interior edge; see Fig. 2.15. We can now define our vertical decomposition in terms of the local divergence free face spaces $\tilde{V}_{l;F}$ and the element spaces $\tilde{V}_{l;T}$. We find the following direct sum representation for $\tilde{V}_l$

$$\tilde{V}_l = \sum_{F \in \mathcal{F}_{l-1} \setminus \mathcal{F}_l} \tilde{V}_{l;F} \oplus \sum_{T \in \mathcal{T}_{l-1} \setminus \mathcal{T}_l} \tilde{V}_{l;T}\ . \tag{2.32}$$

To define our quasi-projection operators, we have to specify a new bilinear form $\tilde{a}(\cdot, \cdot)$ on $\tilde{V}_l \times \tilde{V}_l$. It is given by means of the direct splitting (2.32) and the original bilinear form $a(\cdot, \cdot)$

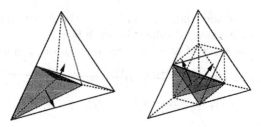

Fig. 2.15. Adaptive refinement such that $\widetilde{V}_{l;T}$ contains a divergence free element

$$\tilde{a}(\mathbf{v}_l, \mathbf{v}_l) := \sum_{F \in \mathcal{F}_{l-1} \setminus \mathcal{F}_l} a(\mathbf{v}_F, \mathbf{v}_F) + \sum_{T \in \mathcal{T}_{l-1} \setminus \mathcal{T}_l} a(\mathbf{v}_T, \mathbf{v}_T) \ ,$$

where according to (2.32) $\mathbf{v}_l := \sum_{F \in \mathcal{F}_{l-1} \setminus \mathcal{F}_l} \mathbf{v}_F + \sum_{T \in \mathcal{T}_{l-1} \setminus \mathcal{T}_l} \mathbf{v}_T.$

The following lemma shows the equivalence of the bilinear forms $\tilde{a}(\cdot, \cdot)$ and $a(\cdot, \cdot)$ when restricted to $\widetilde{V}_l \times \widetilde{V}_l$.

Lemma 2.12. *There exist constants c and C such that*

$$c\,a(\mathbf{v}_l, \mathbf{v}_l) \leq \tilde{a}(\mathbf{v}_l, \mathbf{v}_l) \leq C\,a(\mathbf{v}_l, \mathbf{v}_l), \quad \mathbf{v}_l \in \widetilde{V}_l \ .$$

Proof. We start by proving the lower bound. It can be obtained easily by a coloring argument. We define $\tilde{\mathbf{v}}_F := \sum_{F \in \mathcal{F}_{l-1} \setminus \mathcal{F}_l} \mathbf{v}_F$, $\tilde{\mathbf{v}}_T := \sum_{T \in \mathcal{T}_{l-1} \setminus \mathcal{T}_l} \mathbf{v}_T$, and find

$$a(\mathbf{v}_l, \mathbf{v}_l) = a(\tilde{\mathbf{v}}_F + \tilde{\mathbf{v}}_T, \tilde{\mathbf{v}}_F + \tilde{\mathbf{v}}_T) \leq \tfrac{5}{4} a(\tilde{\mathbf{v}}_F, \tilde{\mathbf{v}}_F) + 5 a(\tilde{\mathbf{v}}_T, \tilde{\mathbf{v}}_T)$$

$$\leq 5 \Big(\sum_{F \in \mathcal{F}_{l-1} \setminus \mathcal{F}_l} a(\mathbf{v}_F, \mathbf{v}_F) + \sum_{T \in \mathcal{T}_{l-1} \setminus \mathcal{T}_l} a(\mathbf{v}_T, \mathbf{v}_T) \Big) = 5\,\tilde{a}(\mathbf{v}_l, \mathbf{v}_l) \ .$$

The proof of the upper bound relies on the norm equivalence (2.10). Let $\mathbf{v}_F$ be an element in $V_{l;F}$, then $a(\mathbf{v}_F, \mathbf{v}_F)^{1/2}$ is equivalent to its L^2-norm:

$$\sum_{F \in \mathcal{F}_{l-1} \setminus \mathcal{F}_l} a(\mathbf{v}_F, \mathbf{v}_F) \leq C \sum_{F \in \mathcal{F}_{l-1} \setminus \mathcal{F}_l} \sum_{f \in \mathcal{F}_l} |f|^{\frac{1}{2}} \|\mathbf{v}_F \cdot \mathbf{n}\|_{0;f}^2$$

$$\leq C \sum_{F \in \mathcal{F}_{l-1} \setminus \mathcal{F}_l} \sum_{\substack{f \in \mathcal{F}_l \\ f \subset F}} |f|^{\frac{1}{2}} \|\mathbf{v}_F \cdot \mathbf{n}\|_{0;f}^2$$

$$= C \sum_{F \in \mathcal{F}_{l-1} \setminus \mathcal{F}_l} \sum_{\substack{f \in \mathcal{F}_l \\ f \subset F}} |f|^{\frac{1}{2}} \|\mathbf{v} \cdot \mathbf{n}\|_{0;f}^2$$

$$\leq C \sum_{f \in \mathcal{F}_l} |f|^{\frac{1}{2}} \|\mathbf{v} \cdot \mathbf{n}\|_{0;f}^2 \leq C a(\mathbf{v}, \mathbf{v}) \ .$$

By means of the triangle inequality, the following upper bound for $a(\mathbf{v}_T, \mathbf{v}_T)$ can easily be established

$$\sum_{T \in \mathcal{T}_{l-1} \setminus \mathcal{T}_l} a(\mathbf{v}_T, \mathbf{v}_T) = a\Big(\mathbf{v} - \sum_{F \in \mathcal{F}_{l-1} \setminus \mathcal{F}_l} \mathbf{v}_F, \mathbf{v} - \sum_{F \in \mathcal{F}_{l-1} \setminus \mathcal{F}_l} \mathbf{v}_F\Big) \leq C a(\mathbf{v}, \mathbf{v}) \ .$$

$\square$

We define now our hierarchical basis method in terms of the decomposition (2.30), the quasi-projections T_l, $1 \leq l \leq j$, given by

$$\tilde{a}(T_l \mathbf{v}, \mathbf{v}_l) := a(\mathbf{v}, \mathbf{v}_l), \quad \mathbf{v}_l \in \widetilde{V}_l$$

and the exact projection T_0 onto V_0 with respect to the bilinear form $a(\cdot, \cdot)$. Then, the central result of this subsection follows from Lemma 2.1, (2.31), Lemma 2.11, and Lemma 2.12. The hierarchical basis method yields a quasi-optimal preconditioner.

Theorem 2.13. *The condition number of the additive Schwarz method is bounded by*

$$\kappa(T_{add}) \leq C(1+j)^2 \ .$$

The exact projection has to be applied on the coarsest level, but on the finer ones we have to solve only local subproblems. For each face, $F \in \mathcal{F}_{l-1} \setminus \mathcal{F}_l$, we have to solve a problem of dimension at most three, and for each element, $T \in \mathcal{T}_{l-1} \setminus \mathcal{T}_l$, a problem of dimension at most eight. The computational effort can be further reduced. The stiffness matrix associated with one face is equivalent to a mass matrix and therefore can be replaced by its diagonal. This is, in general, not the case for the stiffness matrices associated with $\widetilde{V}_{l;T}$ if the nodal basis vector fields are used. However, if a basis transformation is performed, and the divergence free element is spanned by one basis function, we can work with the diagonal of the stiffness matrix.

Remark 2.14. *We note that the theoretical results of Subsect. 2.2.2 also hold for simplicial triangulations and the ones of Subsect. 2.2.3 for hexahedral triangulations.*

2.2.4 Numerical Results

In this subsection, we report on numerical results illustrating the performance of the proposed iterative substructuring method. Although the analysis was carried out only for the 3D case in this chapter, we also include some results for the 2D case. The results for the 2D case can be found in [TWW00], and those for the 3D case in [WTW00]. For further numerical results including two-level overlapping and Neumann–Neumann methods, we refer to [Tos99].

We use the iterative substructuring method defined by the decomposition (2.11) and vary the meshsize of our uniform coarse and fine triangulations. The fine triangulation $\mathcal{T}_h$ consists of n^2 elements, with $h = 1/n$. We observe the theoretically predicted logarithmic behavior of the condition number. In a second test case, we consider the influence of the coefficients a and B, where the matrix B is given by $B := b\,\mathrm{Id}$. We refer to [SBG96], for a general discussion of practical issues concerning Schwarz methods.

2.2.4.1 The 2D Case. We start with the case $a = b = 1$. The following table shows the estimated condition number of the additive Schwarz operator as a function of the dimension of the fine and coarse meshes. In addition, the number of conjugate gradient iterations required to obtain a reduction of the residual norm by a factor of 10^{-6} is given. The columns of Table 2.1 provide the condition numbers for fixed ratios of the meshsizes of the coarse and fine triangulations.

Table 2.1. Condition number and number of cg-iterations, (in parentheses)

H/h	32	16	8	4	2
n=32	-	20.23 (11)	26.50 (20)	19.10 (20)	12.86 (17)
n=64	26.27 (11)	35.94 (20)	27.16 (21)	19.00 (17)	12.90 (16)
n=128	46.83 (20)	36.68 (18)	27.06 (17)	18.92 (16)	x
n=192	-	36.71 (17)	27.00 (17)	18.90 (16)	x
n=256	47.80 (18)	36.66 (17)	26.97 (16)	18.89 (16)	x

For a fixed ratio H/h, the condition number is quite insensitive to the dimension of the fine mesh. The number of iterations varies slowly with H/h and our results compare well with those for finite element approximations of the Laplace equation; see, e.g., [SBG96].

We find in Table 2.1 two values which are considerably smaller than the others in the same column. This is the case of $n = 64$, $H/h = 32$ and $n = 32$, $H/h = 16$ which correspond to a partition into 2 by 2 subregions. The reason for the smaller condition numbers of T_{add} is a smaller norm of the additive Schwarz operator in those cases. We recall that this bound is obtained by a coloring argument. The largest eigenvalue of T_{add} is bounded by 5 in all the cases in Table 2.1, except for $n = 32$, $H/h = 16$ and $n = 64$, $H/h = 32$, when it is bounded by 3.

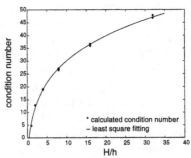

Fig. 2.16. Condition number (asterisk) and least-square second order logarithmic polynomial (solid line)

In Fig. 2.16, we plot the results of Table 2.1 together with the best second order logarithmic polynomial least-square fit. The relative fitting error is about 1.8%. Our numerical results are therefore in good agreement with our theoretical bound and also confirm that our bound is sharp.

Table 2.2. Condition number and number of cg-iterations, (in parentheses)

H/h	32	16	8	4
b=0.00001	3.87 (10)	4.68 (13)	4.86 (13)	4.92 (13)
b=0.0001	3.87 (10)	36.3 (16)	26.2 (16)	13 (15)
b=0.001	16.9 (11)	36.5 (16)	27 (16)	18.7 (16)
b=0.01	46.9 (14)	36.7 (17)	27.1 (16)	18.9 (16)
b=0.1	46.9 (14)	36.7 (17)	27.1 (17)	18.9 (16)
b=1	46.8 (20)	36.7 (18)	27.1 (17)	18.9 (16)
b= 10	45.3 (22)	36.4 (22)	27 (18)	18.9 (17)
b=100	40.8 (25)	34.8 (23)	26.7 (20)	18.9 (19)
b=1000	29.8 (24)	28.4 (23)	24.5 (21)	18.4 (19)
b=10000	17.4 (18)	17.3 (17)	16.8 (18)	15.3 (17)
b=100000	9.41 (14)	9.37 (14)	9.3 (14)	9.15 (14)

In Table 2.2, we show some results when the ratio of the coefficients b and a is changed. The estimated condition number and the number of iterations are shown as a function of H/h and b, for a fixed value of $n = 128$ and $a = 1$. We observe that the condition number tends to be independent of the ratio H/h when the ratio b/a is very small or very large. This behavior is not fully covered by the theory, but an analysis of the limit cases $b = 0$ and $a = 0$ is carried out in [TWW00]; it can be shown that there is a uniformly stable decomposition for $v \in V_h$ for these two limit cases. In particular, the L^2-stability of our decomposition is guaranteed.

We recall that in 2D, the qualitative and quantitative results for our iterative substructuring method for the lowest order Raviart–Thomas finite elements are exactly the same as for Maxwell's equation discretized by Nédélec finite elements. We remark that when Maxwell's equations are discretized with an implicit time-scheme, the time step is related to the ratio b/a which is in general very large. Our iterative substructuring method therefore appears quite attractive for the solution of linear systems arising from the finite element approximation of time-dependent Maxwell's equations.

2.2.4.2 The 3D Case. In this subsubsection, we present the same type of numerical results for the 3D case. The condition numbers for the additive Schwarz method are given for different diameters of the coarse and fine meshes, and coefficients a and b.

In Table 2.3, we show the estimated condition number and the number of conjugate gradient iterations in order to obtain a reduction of the residual norm by a factor 10^{-6}, as a function of the dimensions of the fine and coarse meshes. The condition number remains bounded independently of n for a fixed ratio H/h. It is slowly increasing with H/h. We observe that the number of iterations varies slowly with H/h and n. The largest eigenvalue is bounded by 7 in all the cases in Table 2.3, except for $n = 8$, $H/h = 4$ and $n = 16$, $H/h = 8$; the latter cases correspond to a partition into 2 by 2 by 2 subregions and, consequently, the bound for the largest eigenvalue is 4.

Table 2.3. Condition number and number of cg-iterations, (in parentheses)

H/h	8	4	2
n=8	–	13.28 (14)	15.15 (22)
n=16	19.46 (16)	23.26 (24)	17.37 (24)
n=24	32.78 (27)	25.55 (26)	17.43 (21)
n=32	33.48 (27)	26.01 (26)	17.42 (21)
n=40	35.50 (27)	26.08 (25)	–
n=48	36.47 (28)	25.91 (22)	–

In Fig. 2.17, we plot the results of Table 2.3 together with the best second order logarithmic polynomial least-square fit. The relative fitting error is about 4.4%. Our numerical results are therefore in good agreement with our theoretical bound. Moreover, we observe that the given theoretical bound is sharp. A more careful analysis shows that an example can be explicitely constructed such that the bound is sharp.

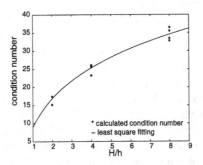

Fig. 2.17. Condition number (asterisk) and least-square second order logarithmic polynomial (solid line)

In Table 2.4, we show some results when the ratio of the coefficients b and a changes. For a fixed value of $n = 24$ and $a = 1$, the estimated condition number and the number of iterations are shown as functions of H/h and b. The condition number tends to be bounded independently of the ratio H/h when the ratio b/a is very small or very large. Generally, the numerical results for the 2D case and the 3D case show the same qualitative structure.

Table 2.4. Condition number and number of cg-iterations, (in parentheses)

H/h	8	4	2
b=1e-09	4.00 (10)	5.81 (16)	6.29 (15)
b=1e-08	4.00 (10)	5.81 (16)	6.29 (15)
b=1e-07	4.00 (10)	5.82 (16)	6.29 (15)
b=1e-06	4.00 (10)	21.0 (18)	6.29 (15)
b=1e-05	17.5 (11)	25.0 (19)	16.1 (18)
b=0.0001	29.5 (12)	25.0 (19)	17.1 (18)
b=0.001	30.9 (15)	25.3 (21)	17.2 (18)
b=0.01	32.3 (20)	25.4 (22)	17.2 (18)
b=0.1	32.6 (22)	25.5 (25)	17.4 (20)
b= 1	32.8 (27)	25.6 (26)	17.4 (21)
b= 10	30.0 (29)	23.4 (26)	17.1 (23)
b=1e+02	23.6 (26)	20.4 (25)	15.1 (22)
b=1e+03	14.4 (21)	14.1 (22)	12.6 (19)
b=1e+04	8.42 (16)	8.57 (17)	9.43 (17)
b=1e+05	6.75 (14)	6.98 (15)	7.92 (17)
b=1e+06	6.72 (14)	6.91 (15)	7.80 (16)

2.3 A Multigrid Method for the Mortar Product Space Formulation

In this section, we consider a multigrid method for mortar finite element discretizations. Recently, a lot of work has been done on efficient iterative solvers for the linear systems that arise from these discretizations; see, e.g., [AK95, AKP95, AMW96, AMW99, BD98, BDL99, BDH99b, BDW99, CW96, Dry98a, Dry98b, EHI+98, EHI+00, GP00, HIK+98, Kuz95a, Kuz95b, Kuz98, KW95, Lac98, Le 93, LSV94, WW98, WW99]. All these techniques are either based on the positive definite variational problem on the constrained nonconforming space or on the equivalent saddle point problem. In the first case, the mortar projection typically has to be applied in each iteration or even in each smoothing step, and this requires the exact solution of a mass matrix system. In the saddle point formulation, a Schur complement system plays an essential role.

Efficient iterative solvers for the nonconforming formulation (1.3) or an equivalent interface problem have been developed. Additive Schwarz methods including iterative substructuring, overlapping and Neumann–Neumann type algorithms as well as multigrid methods have been discussed and analyzed; see, e.g., [BDH99b, CDS98, Dry96, Dry97, Dry98a, Dry98b, GP00, Le 93, LSV94].

Here, we combine the idea of dual basis functions for the Lagrange multiplier space with standard multigrid techniques for symmetric positive definite systems. The new mortar formulation, analyzed in Sect. 1.3, is the point of departure for our multigrid method. This approach was originally studied in [WK99] and has been extended to linear elasticity problems in [KW00b].

In particular, we do not have to solve a modified Schur complement system exactly or iteratively, nor do we have to solve a mass matrix system in each smoothing step.

Here, we assume that $\mathcal{T}_0, \mathcal{T}_1, \ldots, \mathcal{T}_j$ form a nested family of quasi-uniform and shape regular triangulations satisfying $h_{l-1} = 2h_l$, $1 \leq l \leq j$. The associated unconstrained product spaces, denoted by X_l, are then nested. Furthermore, we assume full H^2-regularity of the elliptic problem. We present our multigrid analysis in two different forms. In this section, we work with a matrix setting and choose scaled nodal basis functions for the vector representation. The multigrid analysis in Sect. 2.5 is given in the abstract operator notation. Here, we focus on a multigrid method for mortar finite elements using dual basis spaces for the definition of the Lagrange multiplier. We use neither the nonconforming formulation, (1.14), nor the saddle point formulation, (1.15), but instead the positive definite one, (1.43), on the product space introduced in Sect. 1.3. The starting point of our analysis is the modified system on the product space.

The main difficulty in solving the linear systems arising from mortar discretizations is the handling of the constraints at the interfaces. Using the approach presented in Sect. 1.3, multigrid methods can be applied directly to the positive definite formulation (1.43). In contrast to the constrained nonconforming case, the unconstrained product spaces associated with a nested sequence of triangulations are nested. However, using the standard injection to define the restriction and prolongation would not provide a suitable approximation property.

We define level dependent bilinear forms and suitable transfer operators in Subsect. 2.3.1. By means of the a priori estimates (1.22) and the definition of the level dependent bilinear form, we establish an approximation property in Subsect. 2.3.2. In Subsect. 2.3.3, we consider a general class of smoothing operators for which smoothing and stability properties will be shown. Subsect. 2.3.4 concerns a simplified implementation of the smoother. Finally in Subsect. 2.3.5, numerical results illustrate the performances of $\mathcal{V}$- and $\mathcal{W}$-cycles.

2.3.1 Bilinear Forms

We set $\dim X_l =: n_l$ and denote the Euclidean scalar product in $\mathbb{R}^{n_l}$ by $(\cdot, \cdot)$. Furthermore, we use the same notation for $v_l \in X_l$ and its vector representation $v_l \in \mathbb{R}^{n_l}$ with respect to the standard nodal basis in 2D and the by $h_l^{-1/2}$ scaled nodal basis in 3D. Using these nodal basis functions and this notation yield that the L^2-norm of $v_l \in X_l$ is equivalent in 2D and in 3D to h_l times the Euclidean vector norm, $\|v_l\| := (v_l, v_l)^{1/2}$, of $v_l \in \mathbb{R}^{n_l}$

$$c\,\|v_l\|_0 \leq h_l\,\|v_l\| \leq C\,\|v_l\|_0 \ .$$

Since, we are working with the variational form on the unconstrained product space no norm has to be specified for the Lagrange multiplier.

We recall that the linear form $g_2(\cdot)$ defined in (1.41) is given in terms of the bilinear form $b(\cdot,\cdot)$, the Lagrange multiplier nodal basis functions, and a trivial extension by zero. The stability constant of this extension depends on the meshsize. As a result, the linear form $g_2(\cdot)$ depends on the level l. We define our level dependent bilinear form $\hat{a}_l(\cdot,\cdot)$ in terms of the original bilinear form $a(\cdot,\cdot)$ and $g_2(\cdot)$ by

$$\hat{a}_l(w_l, v_l) := a(w_l - g_2(w_l), v_l - g_2(v_l)) + a(g_2(w_l), g_2(v_l)), \qquad w_l, v_l \in X_l \ .$$

Using Proposition 1.15 and an inverse estimate, it is easy to see that an upper bound for $\hat{a}_l(v,v)$ is given by the square of a mesh dependent norm

$$\hat{a}_l(v, v) \le C \left(\|v\|_1^2 + \frac{1}{h_l} \|[v]\|_{0;\mathcal{S}}^2 \right), \quad v \in \prod_{k=1}^{K} H^1(\Omega_k) \ . \tag{2.33}$$

We observe that $v - g_2(v) \in Y$. Then, the uniform ellipticity of $a(\cdot,\cdot)$ on $Y \times Y$ yields a lower bound for $\hat{a}_l(v,v)$

$$\hat{a}_l(v, v) \ge C(\|v - g_2(v)\|_1^2 + \|g_2(v)\|_1^2) \ge C\|v\|_1^2, \quad v \in \prod_{k=1}^{K} H^1(\Omega_k) \ . \tag{2.34}$$

Working with the bilinear form $\hat{a}_j(\cdot,\cdot)$ on $X_l \times X_l$, $0 \le l \le j$, we cannot expect an approximation property with a constant growing slower than 2^{j-l}. Instead of using the bilinear form $\hat{a}_j(\cdot,\cdot)$ on all levels, we consider the bilinear form $\hat{a}_l(\cdot,\cdot)$ on level l. The stiffness matrix $\hat{A}_l$ is associated with this bilinear form, and the variational problem (1.43) can be written as

$$\hat{A}_l u_l = f_l \ ,$$

on level l. In comparison with formula (1.48), we have added the index l to indicate the level dependency. In the next subsection, we show that a suitable approximation property holds for the finite element spaces X_l if the level dependent bilinear forms $\hat{a}_l(\cdot,\cdot)$ are used and if the defect has a special structure.

To define our multigrid method, we need to specify restriction and prolongation matrices. Let us first consider the standard grid transfer operators to see why they do not meet our requirements. They are denoted by $I_l^{l-1}: \mathbb{R}^{n_l} \longrightarrow \mathbb{R}^{n_{l-1}}$, and by $I_{l-1}^l: \mathbb{R}^{n_{l-1}} \longrightarrow \mathbb{R}^{n_l}$. As usual, I_{l-1}^l is chosen as the matrix representation of the natural injection of X_{l-1} in X_l with respect to the specified nodal basis, and the restriction I_l^{l-1} is defined by $I_l^{l-1} := (I_{l-1}^l)^T$. Thus, we have

$$(I_l^{l-1} v_l, w_{l-1}) = (v_l, I_{l-1}^l w_{l-1}), \qquad v_l \in \mathbb{R}^{n_l}, \quad w_{l-1} \in \mathbb{R}^{n_{l-1}} \ .$$

Let $\tilde{w}_{l-1}$ be the solution of the defect problem in X_{l-1}, i.e., $\hat{A}_{l-1}\tilde{w}_{l-1} = \tilde{d}_{l-1}$, and let $\tilde{d}_{l-1} := I_l^{l-1} d_l \in X_{l-1}$. Then, even if $W_l^T d_l = 0$, in general

$W_{l-1}^T \tilde{d}_{l-1} \neq 0$. We recall that W_l scales the values on the non-mortar sides; see Subsect. 1.3.3. As a consequence, the constraints are in general not satisfied on level $l-1$, i.e., $B_{l-1}^T \tilde{w}_{l-1} \neq 0$. In the next subsection, we show our approximation property only for the special case $W_{l-1}^T d_{l-1} = 0$, and thus we cannot use I_l^{l-1} as a restriction operator. These preliminary remarks show a need for a modified restriction operator $(\tilde{I}_{\mathrm{mod}})_l^{l-1}$ which should satisfy $W_{l-1}^T d_{l-1} = 0$ if $W_l^T d_l = 0$. Our new transfer operators are defined in terms of the local projection

$$P_l := B_l W_l^T \tag{2.35}$$

and are based on the decomposition $\mathrm{Id} = (\mathrm{Id} - P_l) + P_l$. The support of $P_l v_l$, $v_l \in X_l$, is contained in the union of small strips of width h_l located on the non-mortar sides of the interfaces; see Fig. 2.18. We note that $W_l^T(\mathrm{Id} - P_l) = 0$, and we define

$$(\tilde{I}_{\mathrm{mod}})_l^{l-1} := (\mathrm{Id} - P_{l-1}) I_l^{l-1} (\mathrm{Id} - P_l) + P_{l-1} I_l^{l-1} P_l$$

and $(\tilde{I}_{\mathrm{mod}})_{l-1}^l := ((\tilde{I}_{\mathrm{mod}})_l^{l-1})^T$. The coarse grid correction w_{l-1} is now given as the solution of

$$\hat{a}_{l-1}(w_{l-1}, v_{l-1}) = (d_{l-1}, v_{l-1})_0, \qquad v_{l-1} \in X_{l-1} , \tag{2.36}$$

where $d_{l-1} := (\tilde{I}_{\mathrm{mod}})_l^{l-1} d_l$, and d_l is the residual on level l. Thus, if the iterates on level l satisfy the constraints, the correction w_{l-1} will satisfy the constraints on level $l-1$, i.e., $B_{l-1}^T w_{l-1} = 0$.

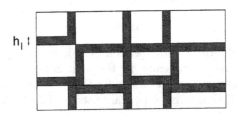

Fig. 2.18. Structure of the support of $P_l v_l$

In the case that the residual d_l on level l satisfies $W_l^T d_l = 0$, we can compute d_{l-1} by $d_{l-1} = (I_{\mathrm{mod}})_l^{l-1} d_l$, where $(I_{\mathrm{mod}})_l^{l-1}$ is the simplified restriction operator

$$(I_{\mathrm{mod}})_l^{l-1} := (\mathrm{Id} - P_{l-1}) I_l^{l-1} . \tag{2.37}$$

If the correction w_{l-1} on level $l-1$ satisfies $B_{l-1}^T w_{l-1} = 0$, the prolongation operator $(\tilde{I}_{\mathrm{mod}})_{l-1}^l$ can be replaced by its simplified form:

$$(I_{\mathrm{mod}})_{l-1}^l := (\mathrm{Id} - P_l^T) I_{l-1}^l . \tag{2.38}$$

Proposition 2.15. *Under the assumption $B_{l-1}^T w_{l-1} = 0$, we have the following norm equivalence*

$$c \, \|w_{l-1}\| \leq \|(I_{\mathrm{mod}})_{l-1}^l w_{l-1}\| \leq C \, \|w_{l-1}\| \,.$$

Proof. The upper bound follows from $\|P_{l-1}\| \leq C$ and the stability of I_{l-1}^l. Observing that $\mathrm{Id} - P_l^T$ modifies only the nodal values at the vertices in the interior of the non-mortar sides, we find $\|(I_{\mathrm{mod}})_{l-1}^l w_{l-1}\| \geq \|I_{l-1}^l w_{l-1}\|_R \geq \|w_{l-1}\|_R$ where $\| \cdot \|_R$ denotes the Euclidean norm without the components associated with the interior of the non-mortar sides. Using the assumption $B_{l-1}^T w_{l-1} = 0$, we find that $\|w_{l-1}\| \leq C \, \|w_{l-1}\|_R$. $\square$

As we will see in Subsect. 2.3.3, our special class of smoothers satisfies $W_l^T d_l = 0$ and $B_{l-1}^T w_{l-1} = 0$ by construction, and we can therefore use $(I_{\mathrm{mod}})_l^{l-1}$ and $(I_{\mathrm{mod}})_{l-1}^l$ in the implementation. Our multigrid method is then defined in terms of the level dependent bilinear form $\hat{a}_l(\cdot, \cdot)$ and the simplified transfer operators. In the next two subsections, we establish suitable approximation and smoothing properties.

2.3.2 An Approximation Property

An essential tool for establishing level independent convergence rates for the $\mathcal{W}$-cycle is a suitable approximation property. The saddle point problem (1.15) forms the starting point of our analysis. We establish an approximation property in the L^2-norm, based on the a priori estimates for the saddle point problem (1.15). In [Woh00b], an approximation property for the saddle point problem on level l was shown if the right hand side is orthogonal to the space $X_{l-1} \times M_{l-1}$ with respect to a level dependent bilinear form. This result was established for the standard Lagrange multiplier spaces which are nested. We cannot apply this result in our setting since the Lagrange multiplier spaces do not have to be nested; in particular, the examples satisfying (1.32) presented in Subsect. 1.2.4 are non-nested, i.e., $M_{l-1} \not\subset M_l$. Instead, we establish a weaker result for the more general situation of non-nested Lagrange multiplier spaces. In the case of standard Lagrange multiplier spaces and the standard prolongation, we refer to [BD98, BDW99].

Let w_l and w_{l-1} be the solution of $\hat{A}_l w_l = d_l$ and $\hat{A}_{l-1} w_{l-1} = d_{l-1}$, respectively, where the stiffness matrices and d_{l-1} are defined in the previous subsection. The following lemma gives an upper bound for $w_l - (I_{\mathrm{mod}})_{l-1}^l w_{l-1}$ in the Euclidean vector norm.

Lemma 2.16. (Approximation property)
Under the assumption that $W_l^T d_l = 0$, the following upper bounds hold

$$\|w_l - I_{l-1}^l w_{l-1}\| \leq C \, \|d_l\| \,,$$

$$\|w_l - (I_{\mathrm{mod}})_{l-1}^l w_{l-1}\| \leq C \, \|d_l\|$$

with a constant C independent of the refinement level l.

Proof. We start with the first inequality. Observing that $W_{l-1}^T d_{l-1} = W_{l-1}^T (\mathrm{Id} - P_{l-1}) I_l^{l-1} d_l = 0$, we find that $B_l^T w_l = B_{l-1}^T w_{l-1} = 0$. To prove the assertion, we consider related saddle point problems. Defining $\lambda_l := W_l^T (d_l - A_l w_l)$ and $\lambda_{l-1} := W_{l-1}^T (I_l^{l-1} d_l - A_{l-1} w_{l-1})$, we find that (w_l, λ_l) and (w_{l-1}, λ_{l-1}) solve the saddle point problems

$$\begin{pmatrix} A_l & B_l \\ B_l^T & 0 \end{pmatrix} \begin{pmatrix} w_l \\ \lambda_l \end{pmatrix} = \begin{pmatrix} d_l \\ 0 \end{pmatrix} \quad \text{and} \quad \begin{pmatrix} A_{l-1} & B_{l-1} \\ B_{l-1}^T & 0 \end{pmatrix} \begin{pmatrix} w_{l-1} \\ \lambda_{l-1} \end{pmatrix} = \begin{pmatrix} I_l^{l-1} d_l \\ 0 \end{pmatrix} ,$$

respectively. We note that the right hand side of the saddle point problem on level $l-1$ is defined in terms of the standard restriction I_l^{l-1}. Associated with d_l is a unique $f_d \in X_l \subset L^2(\Omega)$ such that for $v_l \in X_l$

$$(f_d, v_l)_0 = (d_l, v_l) ,$$

and such that $h_l \|f_d\|_0 \leq C \|d_l\|$. We now define a continuous variational problem in terms of f_d by: Find $w \in H_0^1(\Omega)$ such that

$$a(w, v) = (f_d, v)_0, \quad v \in H_0^1(\Omega) .$$

Then, $w_l \in V_l$ and $w_{l-1} \in V_{l-1}$ are the corresponding nonconforming mortar finite element approximations of this variational problem on level l and $l-1$, respectively. Using the full H^2-regularity and the a priori estimates in the L^2-norm, we obtain

$$\|w_l - I_{l-1}^l w_{l-1}\| \leq \tfrac{C}{h_l} \|w_l - w_{l-1}\|_0 \leq \tfrac{C}{h_l} (\|w_l - w\|_0 + \|w - w_{l-1}\|_0)$$
$$\leq C h_l \|w\|_2 \leq C h_l \|f_d\|_0 \leq C \|d_l\| .$$

The second assertion is based on the triangle inequality

$$\|w_l - (I_{\mathrm{mod}})_{l-1}^l w_{l-1}\| \leq \|w_l - I_{l-1}^l w_{l-1}\| + \|P_l^T I_{l-1}^l w_{l-1}\| .$$

To estimate the second term, we observe that P_l^T is the algebraic representation of $g_2(\cdot)$ on level l. Using the definition of the linear form $g_2(\cdot)$, we find in terms of Lemma 1.7, Proposition 1.15 and the a priori estimates for the energy norm

$$\|P_l^T I_{l-1}^l w_{l-1}\|^2 \leq \tfrac{C}{h_l^2} \|g_2(w_{l-1})\|_0^2 \leq \tfrac{C}{h_l} \sum_{m=1}^{M} \|[w_{l-1}]\|_{0;\gamma_m}^2$$
$$\leq C \|w_{l-1} - w\|_1^2 \leq C h_l^2 \|w\|_2^2 \leq C \|d_l\|^2 . \qquad \square$$

In the case that the Lagrange multiplier spaces are non-nested, we cannot write $w_l - w_{l-1} \in X_l$ as the solution of a saddle point problem on $X_l \times M_l$. The approximation property then has to be established in terms of $\|w_l - w\|_0$ and $\|w_{l-1} - w\|_0$, where w is a suitable element in $H_0^1(\Omega)$. We point out that the approximation property is only shown for the special case that $W_l^T d_l = 0$. We therefore have to use special types of smoothers to satisfy this condition.

2.3.3 Smoothing and Stability Properties

In addition to the approximation property, a suitable smoothing property has to be established to obtain level independent convergence rates. To measure the smoothing effect, $\|\hat{A}_l e_l^m\|$ has to be bounded by $\|e_l^0\|$, where e_l^m is the iteration error in the mth-step and e_l^0 the initial error. In contrast to the two grid convergence analysis, approximation and smoothing properties are not sufficient to obtain a level independent convergence rate in the full multigrid case. This is so since the stability constant of the smoothing operator in the $\|\cdot\|$-norm will also enter into the estimates.

In this subsection, we construct suitable smoothing operators for the system on the unconstrained product space. In contrast to the other sections, here we use the symbol λ to denote eigenvalues. We focus in particular on a special structure of the smoother yielding iterates which satisfy the constraints at the interface. As a consequence, we obtain a residual satisfying $W_l^T d_l^m = 0, m \geq 1$, for any number of smoothing steps. Only in this case, are we in the setting of Lemma 2.16, which concerns the approximation property.

To define a suitable smoother for our multigrid method, we start with the 2×2 block decomposition of $\hat{A}_l$ introduced in Sect. 1.3. We recall that the first block is obtained by grouping the vectors having index 0, 1, or 2 together. Using $u_R^T := (u_0^T, u_1^T, u_2^T)$, $u_N := u_3$, (1.48) can be rewritten as

$$\begin{pmatrix} \hat{A}_{RR} & \hat{A}_{RN} \\ \hat{A}_{NR} & \hat{A}_{NN} \end{pmatrix} \begin{pmatrix} u_R \\ u_N \end{pmatrix} = \begin{pmatrix} \hat{f}_R \\ 0 \end{pmatrix} \ .$$

For simplicity, we use the level index l only for the global matrix and vector symbols, e.g., the stiffness matrix $\hat{A}_l$ and the solution u_l, but not for subblocks, e.g., $\hat{A}_{RR}$. According to (1.49), the blocks are given by $\hat{A}_{RR} = A_{RR} + 2MA_{NN}M^T - MA_{NR} - A_{RN}M^T$, $\hat{A}_{NR}^T = \hat{A}_{RN} = MA_{NN}$ and $\hat{A}_{NN} = A_{NN}$. Here, M is the scaled mass matrix given in Subsect. 1.3.3 as $M = M_R D^{-1}$. Thus, $\hat{A}_{NN}$ is the symmetric positive definite submatrix of $\hat{A}_l$ associated with the interior degrees of freedom on the non-mortar sides. Its eigenvalues are between c and C.

Proposition 2.17. *The matrix $\hat{A}_{RR}$ is symmetric positive definite with eigenvalues bounded from below by ch_l^2 and from above by a constant C, and $\|\hat{A}_{RR}\| \geq c$.*

Proof. To obtain an upper bound for the eigenvalues of $\hat{A}_{RR}$, we start with the definition of $\hat{a}(\cdot, \cdot)$, observe (2.33) and use an inverse estimate

$$x_R^T \hat{A}_{RR} x_R = \hat{a}(x_R, x_R) \leq C \frac{1}{h_l^2} \|x_R\|_0^2 \leq C \|x_R\|^2 \ .$$

The lower bound is an easy consequence of (2.34), and the bound for the norm can be obtained by choosing one interior nodal basis function as test function in the definition of the norm.   □

Let G_R be a symmetric, positive definite smoothing operator for $\hat{A}_{RR}$, satisfying

$$\sigma(G_R^{-1}\hat{A}_{RR}) \subset [0,1], \quad c\,\mathrm{Id} \leq G_R \leq C\,\mathrm{Id} \ . \tag{2.39}$$

Here, $\sigma(T)$ denotes the spectrum of the operator T. We then define our iteration matrix G_l^{-1} as a 2×2 block matrix by

$$G_l^{-1} := \begin{pmatrix} G_R^{-1} & -G_R^{-1}M \\ -M^T G_R^{-1} & \frac{3}{2}M^T G_R^{-1}M + \alpha\,\mathrm{Id} \end{pmatrix} . \tag{2.40}$$

The coefficient α is defined by $\alpha := 0.5 \cdot \lambda_{\max;N}^{-1}$, where $\lambda_{\max;N}$ is the maximum eigenvalue of $\hat{A}_{NN}$. Now our iteration scheme is given in terms of G_l^{-1}. The mth-iterate $z_l^m \in X_l$ is defined as

$$z_l^m := z_l^{m-1} + G_l^{-1}(d_l - \hat{A}_l z_l^{m-1}), \quad m \geq 1 \ , \tag{2.41}$$

where d_l stands for the right hand side of the system $\hat{A}_l z_l = d_l$ which has to be solved, z_l^m the iterate in the mth-step, and z_l^0 the initial guess. Each smoothing step can easily be performed provided that the application of G_R^{-1} is cheap.

Remark 2.18. *A possible choice for G_R is a damped Richardson method, i.e., $G_R := \mu\,\mathrm{Id}$, where $C \geq \mu \geq \lambda_{\max;R}$, and $\lambda_{max;R}$ is the maximum eigenvalue of $\hat{A}_{RR}$. Then, G_R satisfies (2.39).*

The stability constant of the iteration (2.41) depends essentially on the condition number of the operator G_l^{-1}. The following lemma shows that the algebraic properties of G_l^{-1} are inherited from G_R^{-1}. In particular, the condition number of G_l^{-1} is bounded by that of G_R.

Lemma 2.19. *Under the assumptions (2.39), G_l^{-1} defined in (2.40) is a symmetric and positive definite operator satisfying*

$$c\,\mathrm{Id} \leq G_l \leq C\,\mathrm{Id}$$

with constants c, C independent of the meshsize.

Proof. The symmetry of G_l^{-1} can be seen in formula (2.40). There remains to estimate the eigenvalues of G_l^{-1}. A straightforward computation shows that G_l^{-1} is spectrally equivalent to its block diagonal matrix

$$c_1 x_l^T \left(\mathrm{diag}\ G_l^{-1}\right) x_l \leq x_l^T G_l^{-1} x_l \leq C_1 x_l^T \left(\mathrm{diag}\ G_l^{-1}\right) x_l \ ,$$

where $\mathrm{diag}\ G_l^{-1}$ is the block diagonal matrix of G_l^{-1}, and $c_1 := (5 - \sqrt{17})/4$, $C_1 := (5 + \sqrt{17})/4$. Considering $\mathrm{diag}\ G_l^{-1}$ in more detail, we find by means of $\|M\| \leq C$ and (2.39) the following upper bound

$$x_l^T \left(\text{diag } G_l^{-1}\right) x_l \leq C \, \|G_R^{-1}\| \, \left(\|x_R\|^2 + \|M x_N\|^2 + \alpha \|x_N\|^2\right)$$

$$\leq C \, \|G_R^{-1}\| \, \|x_l\|^2 \;,$$

where $x_l^T := (x_R^T, x_N^T)$. In the first inequality, we have used the fact that $\|G_R\| \leq C$ and in the last one that α is bounded. By using that $\alpha \geq c$ and (2.39), we can give a lower bound for $x_l^T G_l^{-1} x_l$ in terms of $\|G_R\|$

$$x_l^T \left(\text{diag } G_l^{-1}\right) x_l \geq \|G_R\|^{-1} \left(\|x_R\|^2 + \alpha \|x_N\|^2\right) \geq C \|G_R\|^{-1} \|x_l\|^2 \;. \quad \square$$

We remark that, in general, an inequality of the form $\|M x_N\| \geq c \|x_N\|$ with $c > 0$ does not hold. With G_l^{-1} positive definite, we still have to show that G_l provides a good approximation of the high frequency part of $\hat{A}_l$ in order to obtain a smoother. We recall that the modified restriction introduced in Subsect. 2.3.1 guarantees that the right hand side of (2.36) on level $l-1$ has the form $d_{l-1}^T = (d_R^T, 0)$, if the defect on the previous level has this form.

Lemma 2.20. (Smoothing property)
Under the assumptions $B_l^T z_l^0 = 0$, $W_l^T d_l = 0$ and (2.39) on G_R, we obtain

$$\|\hat{A}_l e_l^m\| \leq \frac{C}{m} \|e_l^0\|, \qquad m \geq 1 \;,$$

where $e_l^m := z_l - z_l^m$. Furthermore, each iterate z_l^m satisfies $B_l^T z_l^m = 0$.

Proof. We start by proving $B_l^T z_l^m = 0$ by induction. Assuming that $B_l^T z_l^m = 0$, we find for $B_l^T z_l^{m+1}$

$$B_l^T z_l^{m+1} = B_l^T z_l^m + B_l^T G_l^{-1} \hat{A}_l e_l^m = B_l^T G_l^{-1} \hat{A}_l e_l^m \;.$$

Considering $B_l^T G_l^{-1} \hat{A}_l$ in more detail, and observing that $W_l^T d_l = 0$ gives $B_l^T z_l = 0$, we obtain $M^T e_R^m + e_N^m = D^{-1} B_l^T e_l^m = 0$ and thus

$$B_l^T G_l^{-1} \hat{A}_l e_l^m = B_l^T G_l^{-1} \begin{pmatrix} \hat{A}_{RR} & \hat{A}_{RN} \\ \hat{A}_{NR} & \hat{A}_{NN} \end{pmatrix} \begin{pmatrix} e_R^m \\ e_N^m \end{pmatrix}$$

$$= B_l^T \begin{pmatrix} G_R^{-1} & -G_R^{-1} M \\ -M^T G_R^{-1} & \frac{3}{2} M^T G_R^{-1} M + \alpha \text{Id} \end{pmatrix} \begin{pmatrix} \hat{A}_{RR} e_R^m + \hat{A}_{RN} e_N^m \\ 0 \end{pmatrix}$$

$$= B_l^T \begin{pmatrix} \text{Id} \\ -M^T \end{pmatrix} G_R^{-1} (\hat{A}_{RR} e_R^m + \hat{A}_{RN} e_N^m)$$

$$= \left(M_R^T - D M^T\right) G_R^{-1} \left(\hat{A}_{RR} e_R^m + \hat{A}_{RN} e_N^m\right) = 0 \;.$$

Here, we have used $\hat{A}_{NR} = \hat{A}_{NN} M^T$. Now, the assumption $B_l^T z_l^0 = 0$ yields the assertion $B_l^T z_l^{m+1} = 0$ for $m \geq 1$.

In a second step, we prove the smoothing property of the iteration (2.41). As shown in Lemma 2.19, the operator G_l^{-1} is symmetric and positive definite. Then, $G_l^{-1/2}$ is well defined and $\hat{A}_l e_l^m$ can be rewritten as

$$\hat{A}_l e_l^m = G_l^{1/2}(G_l^{-1/2}\hat{A}_l G_l^{-1/2})(\mathrm{Id} - G_l^{-1/2}\hat{A}_l G_l^{-1/2})^m G_l^{1/2} e_l^0 .$$

Since the norm of a symmetric matrix is bounded by its spectral radius, it is sufficient to estimate the eigenvalues of $G_l^{-1/2}\hat{A}_l G_l^{-1/2}$

$$\|\hat{A}_l e_l^m\| \leq \|G_l\| \max_{s\in\sigma(G_l^{-1}\hat{A}_l)} |s(1-s)^m| \, \|e_l^0\| .$$

We therefore consider the eigenvalue problem $G_l^{-1}\hat{A}_l x_l = \lambda \, x_l$, $x_l^T := (x_R^T, x_N^T)$, and prove $0 \leq \lambda \leq 1$. To start, we give the block structure of $G_l^{-1}\hat{A}_l$

$$G_l^{-1}\hat{A}_l = \left(\begin{array}{c|c} G_R^{-1}\hat{A}_{RR} - G_R^{-1}M\hat{A}_{NN}M^T & 0 \\ \hline \frac{3}{2}M^T G_R^{-1}M\hat{A}_{NN}M^T & \frac{1}{2}M^T G_R^{-1}M\hat{A}_{NN} \\ +\alpha\hat{A}_{NR} - M^T G_R^{-1}\hat{A}_{RR} & +\alpha\hat{A}_{NN} \end{array}\right) .$$

Here, we have used $\hat{A}_{RN} = M\hat{A}_{NN}$. Obviously, we find that $\lambda > 0$ since G_l^{-1} and $\hat{A}_l$ are symmetric and positive definite. Let us first assume $x_R \neq 0$. Then since $G_l^{-1}\hat{A}_l$ is a lower block triangular matrix, λ is an eigenvalue of $G_l^{-1}\hat{A}_l$ with the eigenvector x_l only if x_R is an eigenvector of the lower dimensional problem

$$(\hat{A}_{RR} - M\hat{A}_{NN}M^T)x_R = \lambda \, G_R x_R .$$

By using that $x_R^T G_R x_R \geq x_R^T \hat{A}_{RR} x_R$ follows from (2.39) and $\hat{A}_{NN} > 0$, we get

$$x_R^T G_R x_R \geq x_R^T \hat{A}_{RR} x_R - x_R^T M\hat{A}_{NN}M^T x_R = \lambda \, x_R^T G_R x_R ,$$

and thus $\lambda \leq 1$.

Let us now consider the second case $x_l^T = (0, x_N^T)$, i.e., $x_R = 0$. Then, the eigenvalue problem $G_l^{-1}\hat{A}_l x_l = \lambda \, x_l$ reduces to the following eigenvalue problem on a smaller space

$$\left(\frac{1}{2}M^T G_R^{-1}M + \alpha\mathrm{Id}\right) y_N = \lambda \, \hat{A}_{NN}^{-1} y_N, \quad x_N := \hat{A}_{NN}^{-1} y_N .$$

To prove $\lambda \leq 1$, we use that $\hat{A}_l$ is positive definite. We set $w_l^T := ((\hat{A}_{RR}^{-1}M y_N)^T, (-\hat{A}_{NN}^{-1} y_N)^T)$, then $w_l^T \hat{A}_l w_l \geq 0$ yields

$$y_N^T M^T \hat{A}_{RR}^{-1} M y_N \leq y_N^T \hat{A}_{NN}^{-1} y_N .$$

Using the assumption (2.39) on G_R and the definition of α, a straightforward calculation shows

$$\lambda \, y_N^T \hat{A}_{NN}^{-1} y_N = \frac{1}{2}y_N^T M^T G_R^{-1} M y_N + \alpha y_N^T y_N$$

$$\leq \frac{1}{2}y_N^T M^T \hat{A}_{RR}^{-1} M y_N + \frac{1}{2}y_N^T \hat{A}_{NN}^{-1} y_N \leq y_N^T \hat{A}_{NN}^{-1} y_N .$$

Since $\hat{A}_{NN}^{-1} = A_{NN}^{-1}$ is symmetric positive definite, we find that $\lambda \leq 1$. Thus, the eigenvalues of $G_l^{-1/2}\hat{A}_l G_l^{-1/2}$ are bounded from below by zero and from above by one, and by means of Lemma 2.19

$$\|\hat{A}_l e_l^m\| \leq \frac{\|G_l\|}{e\,m}\|e_l^0\| \leq \frac{C}{m}\|e_l^0\| . \qquad \square$$

In contrast to the algorithms given in [BD98, BDW99], there is no need to solve a Schur complement system to guarantee that $B_l^T z_l^m = 0$ in each smoothing step. The action of G_l^{-1} can be obtained easily by applying G_R^{-1}.

Remark 2.21. *The second column of G_l^{-1} is redundant in the application of the smoother and can be replaced by zero. Thus, the implementation of the iteration (2.41) does not require the value of α. Furthermore in Subsect. 2.3.5, we will show that G_l, in (2.41), can be replaced by a Gauß–Seidel smoother.*

In addition to the smoothing and approximation property, the $\|\cdot\|$-stability of the iteration matrix $\mathrm{Id} - G_l^{-1}\hat{A}_l$ is also required for the convergence of the $\mathcal{W}$-cycle; see, e.g., [BS94, Hac85]. The stability is inherited from G_R^{-1} just as the smoothing property.

Lemma 2.22. (Stability estimate)
Under the assumptions (2.39), the iteration (2.41) is stable and the iteration error e_l^m is bounded by

$$\|e_l^m\| \leq C\|e_l^0\|, \quad m \geq 1 .$$

Proof. Since the iteration error e_l^m can be written as

$$e_l^m = (\mathrm{Id} - G_l^{-1}\hat{A}_l)^m e_l^0 ,$$

we can bound $\|e_l^m\|$ by

$$\|e_l^m\| \leq \|G_l^{-1/2}\|\,\|\mathrm{Id} - G_l^{-1/2}\hat{A}_l G_l^{-1/2}\|^m\,\|G_l^{1/2}\|\,\|e_l^0\| .$$

By using that $\sigma(G_l^{-1}\hat{A}_l) \subset [0,1]$, we find that $\|e_l^m\| \leq \sqrt{\kappa(G_l)}\,\|e_l^0\|$, and the stability of the iteration (2.41) follows immediately from Lemma 2.19. $\square$

We can now formulate our main result for the multigrid method defined by the smoothing iteration, (2.41), the coarse grid problem, (2.36), and the modified prolongation $(I_{\mathrm{mod}})_{l-1}^l$ for the coarse grid correction.

Theorem 2.23. *Under the assumptions $B_l^T z_l^0 = 0$ and (2.39), the convergence rates for the $\mathcal{W}$-cycle are independent of the number of refinement levels provided that the number of smoothing steps is large enough.*

Proof. Using our special type of smoother guarantees that $B_l^T z_l^m = 0$ if $B_l^T z_l^0 = 0$. Then the residual d_l satisfies $W_l^T d_l = 0$, and by applying our special restriction operator $(I_{\mathrm{mod}})_l^{l-1}$, we find that $W_{l-1}^T d_{l-1} = 0$. Thus the assumptions of Lemma 2.16 are satisfied, and the proof follows from the approximation property Lemma 2.16, the smoothing property Lemma 2.20, and the stability estimate Lemma 2.22. For details on the general theory, see, e.g., [BS94, Th. 6.5.9] or [Hac85, Th. 7.1.2]. $\square$

2.3.4 Implementation of the Smoothing Step

So far the smoothing step (2.41) is based on G_l^{-1} and $\hat{A}_l$. However due to the special structure of the residual, we can replace G_l by a lower block triangle matrix. In addition, $\hat{A}_l$ does not have to be assembled. Introducing the non-symmetric matrix

$$
A_{\mathrm{non}} = \left(\begin{array}{c|c} \begin{array}{c} A_{RR} + 2M A_{NN} M^T \\ -M A_{NR} - A_{RN} M^T \end{array} & M A_{NN} \\ \hline M^T & \mathrm{Id} \end{array} \right) ,
$$

the smoothing iteration (2.41) can be simplified. For simplicity, we suppress the level index for the matrices if the level l is clear from the context.

Proposition 2.24. *Under the assumptions $B_l^T z_l^0 = 0$ and $W_l^T d_l = 0$ the smoothing iteration*

$$
z_l^m := z_l^{m-1} + \left(\begin{array}{cc} G_R & 0 \\ M^T & \mathrm{Id} \end{array} \right)^{-1} \left(d_l - A_{\mathrm{non}} z_l^{m-1} \right) \tag{2.42}
$$

yields the same results as (2.41).

Proof. A straightforward calculation, as in the proof of Lemma 2.20, shows that $B_l^T z_l^m = 0$ and that the components associated with the index N of the residual are zero. Thus, the action of the non-symmetric smoother in (2.42) is the same as that of the symmetric one given by (2.40) and (2.41).   □

Remark 2.25. *We note that (2.42) can be interpreted as an inexact block Gauß–Seidel smoother for A_{non}. The inverse of $\hat{A}_{RR}$ is replaced by G_R^{-1}.*

A different class of smoothers can be easily constructed by using postprocessing techniques. The point of departure for the introduction of a modified smoothing step is the following observation.

Lemma 2.26. *Let $v_l \in X_l$ be a solution of*

$$
A_{\mathrm{semi}} v_l := (\mathrm{Id} - P_l) A_l (\mathrm{Id} - P_l^T) v_l = (\mathrm{Id} - P_l) f_l .
$$

Then, $u_l := (\mathrm{Id} - P_l^T) v_l$ is the unique solution of (1.48) on level l.

The proof is straightforward.

We remark that A_{semi} is a positive semi-definite matrix. Observing $B_l^T W_l = \mathrm{Id}$, we find that the kernel of A_{semi} is given by the range of W_l. Furthermore, the rows of A_{semi} associated with the interior nodes on the non-mortar side are zero. Our new smoothing iteration is now defined in terms of a good smoother G_{semi}^{-1} for A_{semi}

$$
z_l^m := z_l^{m-1} + (\mathrm{Id} - P_l^T) G_{\mathrm{semi}}^{-1} (\mathrm{Id} - P_l) \left(d_l - A_{\mathrm{semi}} z_l^{m-1} \right) , \tag{2.43}
$$

where G_{semi}^{-1} is a suitable symmetric pseudo-inverse of G_{semi} and satisfies

$$\sigma(G_{\text{semi}}^{-1} A_{\text{semi}}) \subset [0,1], \quad \|G_{\text{semi}}^{-1}\|, \|G_{\text{semi}}\| \le C . \tag{2.44}$$

The following theorem shows that the smoothing step (2.42) can be replaced by (2.43) without losing optimal convergence rates.

Theorem 2.27. *The smoothing iteration (2.43) guarantees level independent convergence rates for the $\mathcal{W}$-cycle provided that the number of smoothing steps is large enough, and the assumptions (2.44) on G_{semi}^{-1} are satisfied and $B_l^T z_l^0 = 0$.*

The proof follows the lines of Subsect. 2.3.3. For details we refer to [KW00a].

We remark that the implementation of the smoothing iteration (2.43) can be based on the positive semi-definite matrix A_{semi} and one local postprocessing step.

Proposition 2.28. *Under the assumptions $z_l^0 := (\mathrm{Id} - P_l^T)\tilde{z}_l^0$ and $W_l^T d_l = 0$, the iterates z_l^m defined by (2.43) can be obtained from $\tilde{z}_l^m$ by the local postprocessing step*

$$z_l^m := (\mathrm{Id} - P_l^T)\tilde{z}_l^m, \quad m \ge 1 .$$

Here, $\tilde{z}_l^m$ is defined as $\tilde{z}_l^m := \tilde{z}_l^{m-1} + G_{\text{semi}}^{-1}(d_l - A_{\text{semi}}\tilde{z}_l^{m-1})$.

Proof. The assertion can be easily shown by induction

$$z_l^m = z_l^{m-1} + (\mathrm{Id} - P_l^T)G_{\text{semi}}^{-1}(d_l - A_{\text{semi}}z_l^{m-1})$$

$$= (\mathrm{Id} - P_l^T)(\tilde{z}_l^{m-1} + G_{\text{semi}}^{-1}(d_l - A_{\text{semi}}\tilde{z}_l^{m-1})) = (\mathrm{Id} - P_l^T)\tilde{z}_l^m . \quad \square$$

2.3.5 Numerical Results in 2D and 3D

We present some numerical experiments illustrating the performance of the algorithm. The theoretical results for the $\mathcal{W}$-cycle are confirmed by our experiments. Furthermore, we show observed level independent convergence rates for the $\mathcal{V}$-cycle. We use piecewise linear Lagrangian finite elements in 2D and the dual space M_h^3, introduced in Subsubsect. 1.2.4.1, as the Lagrange multiplier space. In 3D, we use hexahedral triangulations and the piecewise bilinear simplified dual basis functions ψ_l^s introduced in Subsubsect. 1.2.4.2.

Standard uniform refinement techniques are applied on the triangulations of the different subdomains; each element is decomposed into four congruent subtriangles in 2D and into eight subhexahedras in 3D. The numerical results are based on the proposed multigrid algorithm. In particular, we use the level dependent restriction and prolongation operators given by (2.37) and (2.38), respectively. We compare three different smoothing operators and the influence of the number of smoothing steps on the convergence rates. In the first case, G_R is a damped Jacobi method, $G_R := \mu \, \mathrm{diag} \, A_{RR}$, $\mu = 5/4$. Then, Theorem 2.23 yields a level independent convergence rate for the $\mathcal{W}$-cycle provided that the number of smoothing steps is large enough. Additionally, we apply a symmetric and a non-symmetric Gauß–Seidel smoother for G_R, where the unknowns u_R are ordered lexicographically. The implementation follows Subsect. 2.3.4, and one smoothing step is realized in terms of (2.42). We remark that the ordering in the block Gauß–Seidel method (2.42) is important. Only if the unknowns $u_N = u_3$ are considered after the unknowns u_1 and u_2, will the constraints at the interfaces be satisfied in each smoothing step. To test our method, our multigrid start iterate is set to zero on each refinement level. In practice, nested iteration in terms of the modified prolongation would be used to define the start iterate on the next level. The start iterates w_l^0 for the smoothing steps are defined as usual. They are set to zero in the presmoothing process on level $l < j$ and are defined in terms of the actual iterate and the prolongated defect correction for the postsmoothing step. Now, the definition of our modified prolongation operator yields that $B_l^T w_l^0 = 0$ for all start iterates.

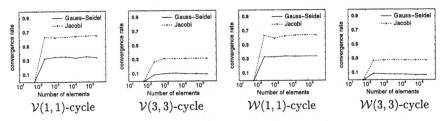

$\mathcal{V}(1,1)$-cycle $\mathcal{V}(3,3)$-cycle $\mathcal{W}(1,1)$-cycle $\mathcal{W}(3,3)$-cycle

Fig. 2.19. Jacobi and symmetric Gauß–Seidel smoother, (Example 1)

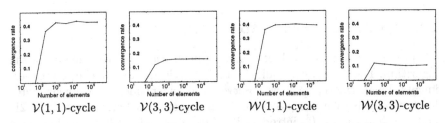

$\mathcal{V}(1,1)$-cycle $\mathcal{V}(3,3)$-cycle $\mathcal{W}(1,1)$-cycle $\mathcal{W}(3,3)$-cycle

Fig. 2.20. Non-symmetric Gauß–Seidel smoother, (Example 1)

Figures 2.19–2.24 show the convergence rates of the $\mathcal{V}$-cycle and the $\mathcal{W}$-cycle for the three different smoothers versus the number of elements. We consider Examples 1–3, introduced in Subsect. 1.5.1. The results for the Jacobi and the symmetric Gauß–Seidel were originally presented in [WK99]. In Figs. 2.19, 2.21 and 2.23, we compare the Jacobi and the symmetric Gauß–Seidel smoother, whereas in Figs. 2.20, 2.22 and 2.24, the non-symmetric

Gauß–Seidel smoother is used. All smoothers can be easily applied, and no extra work is required to satisfy the constraints at the interface. The two pictures on the left in the figures show the results for the $\mathcal{V}$-cycle, whereas the two pictures on the right show the corresponding results for the $\mathcal{W}$-cycle. The number of pre- and postsmoothing steps is equal m; we provide results for $m = 1$ and $m = 3$.

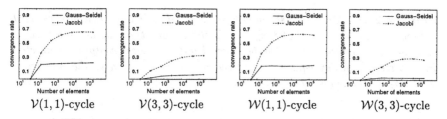

$\mathcal{V}(1,1)$-cycle $\mathcal{V}(3,3)$-cycle $\mathcal{W}(1,1)$-cycle $\mathcal{W}(3,3)$-cycle

Fig. 2.21. Jacobi and symmetric Gauß–Seidel smoother, (Example 2)

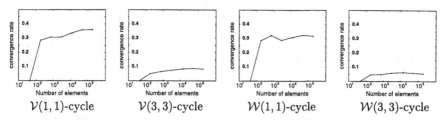

$\mathcal{V}(1,1)$-cycle $\mathcal{V}(3,3)$-cycle $\mathcal{W}(1,1)$-cycle $\mathcal{W}(3,3)$-cycle

Fig. 2.22. Non-symmetric Gauß–Seidel smoother, (Example 2)

We start with a comparison of the numerical results for Examples 1 and 3; see Figs. 2.19, 2.20, 2.23 and 2.24. Here, we have full regularity, and obtain level independent convergence rates for all test settings, even in the case of the $\mathcal{V}$-cycle with just one pre- and one postsmoothing step and as well as for the non-symmetric Gauß–Seidel smoother. Increasing the number of smoothing steps from one to three improves the convergence rates for both the $\mathcal{W}$-cycle and $\mathcal{V}$-cycle considerably. The numerical results show that the convergence rates are approximately three times smaller for three smoothing steps than for one smoothing step. The convergence rates for the $\mathcal{W}$-cycles are only minimally better than those for the $\mathcal{V}$-cycles. The results for the symmetric Gauß–Seidel smoother are better than the results for the Jacobi smoother and the non-symmetric Gauß–Seidel method. However, one application of the symmetric variant is twice as expensive as one of the non-symmetric. The non-symmetric Gauß–Seidel method yields considerably better results than the Jacobi smoother.

Finally, we consider the results for Example 2 in more detail. Figures 2.21 and 2.22 show the observed convergence rates. Although we have no H^2-

regularity, we observe level independent convergence rates; see Figs. 2.21 and
2.22. In the case of the non-symmetric Gauß–Seidel and the Jacobi smoother,
the asymptotic phase starts later than in the other examples. Increasing the
number of smoothing steps gives considerably better results. Using the symmetric Gauß–Seidel variant, we obtain, even with just one smoothing step,
convergence rates where the asymptotic can be observed from the beginning;
see Fig. 2.21.

Remark 2.29. *More robust smoothers, e.g., of ILU type, can also be used,*
if they are modified so that they satisfy the assumptions of Lemma 2.16. A
suitable modification can easily be carried out in a local postprocessing step.
In this step, a scalar equation is solved for each unknown in the interior of
the non-mortar sides; see Proposition 2.28.

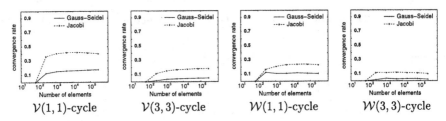

$\mathcal{V}(1,1)$-cycle $\mathcal{V}(3,3)$-cycle $\mathcal{W}(1,1)$-cycle $\mathcal{W}(3,3)$-cycle

Fig. 2.23. Jacobi and symmetric Gauß–Seidel smoother, (Example 3)

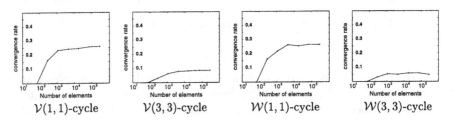

$\mathcal{V}(1,1)$-cycle $\mathcal{V}(3,3)$-cycle $\mathcal{W}(1,1)$-cycle $\mathcal{W}(3,3)$-cycle

Fig. 2.24. Non-symmetric Gauß–Seidel smoother, (Example 3)

We point out that in our approach the constraints can be satisfied in
each smoothing step by applying a simple Gauß–Seidel smoother. Thus, no
additional work is necessary to obtain iterates in the constrained space. The
only difference, compared with a standard multigrid algorithm for symmetric positive definite problems on nested spaces, is the choice of the transfer
operators. In our case, we replace the standard restriction operator I_l^{l-1}
by the level dependent operator $(I_{\mathrm{mod}})_l^{l-1} = (\mathrm{Id} - P_{l-1}) I_l^{l-1}$ as defined in
(2.37), and the modified prolongation given by (2.38) is used. Observing that
P_{l-1} is not only sparse but that the number of non zero entries is bounded
by $C\sqrt{n}$ in 2D and by $Cn^{2/3}$ in 3D, we find that the extra amount of work

is considerably less than one smoothing step; here, n is the number of unknowns. In particular, the modifications involve only a scaled mass matrix on the interface and can be carried out as a local postprocessing step, i.e.,

$$\text{Id} - P_{l-1} = \begin{pmatrix} \text{Id} & 0 \\ M^T & 0 \end{pmatrix} .$$

Finally, we consider two examples in 3D illustrating the performance of our multigrid method. Here, we use trilinear finite elements on hexahedrons and the simplified dual Lagrange multiplier space introduced in Subsubsect. 1.2.4.2. As in the 2D case, we compare the asymptotic convergence rates of the $\mathcal{V}$- and $\mathcal{W}$-cycles in case of one and three smoothing steps. We apply a symmetric Gauß–Seidel and a Jacobi smoother.

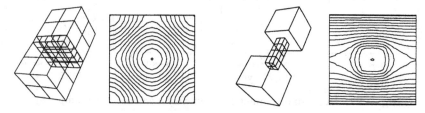

Fig. 2.25. Triangulation and isolines for Example 4 (left) and Example 5 (right)

In Example 4, we consider a sandwich-like composite material. The domain $\overline{\Omega} := \cup_{i=1}^{3} \overline{\Omega}_i$ is decomposed into three hexahedrons $\Omega_i := \{(0,1)^2 \times (z_i, z_{i+1})\}$ where $z_1 := 0$, $z_2 := 1$, $z_3 := 1.2$, $z_4 := 2.2$. As model problem, we consider $-\text{div}\, a\nabla u = 1$ on Ω where the coefficient a is piecewise constant, $a|_{\Omega_i} := 100$, $i = 1, 3$ and $a|_{\Omega_2} := 1$. Dirichlet boundary conditions are applied on the upper and lower part of the domain, $u(x, y, z) = 1000\,((x - 1/2)^2 + (y - 1/2)^2)^{1/2} \cdot (1 - y/3)\exp(-10(x^2 + y^2))$ if $z = z_1$ or $z = z_4$, and homogeneous Neumann boundary conditions are taken elsewhere. In the left part of Fig. 2.25, the nonmatching initial triangulation and the isolines at the interface are shown.

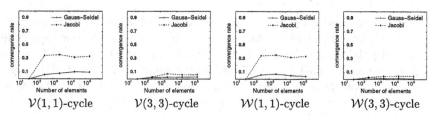

$\mathcal{V}(1,1)$-cycle $\mathcal{V}(3,3)$-cycle $\mathcal{W}(1,1)$-cycle $\mathcal{W}(3,3)$-cycle

Fig. 2.26. Jacobi and symmetric Gauß–Seidel smoother, (Example 4)

The non-mortar sides are defined on the middle hexahedron and the dimension of the Lagrange multiplier space $M_h(\gamma_m)$ is 16 on the initial triangulation. As can be seen in Fig. 2.26, the performance in 3D is comparable to the 2D results. The numerical results confirm the theoretical ones. Asymptotic convergences rates which do not depend on the refinement level can be observed. Even for the $\mathcal{V}(1,1)$-cycle, a constant asymptotic convergence rate is obtained.

In Example 5, we consider a non-convex domain shown in the right part of Fig. 2.25. It is decomposed into three subdomains $\Omega_1 := \{(0,1)^2 \times (0,1)\}$, $\Omega_2 := \{(1/3, 2/3)^2 \times (1,2)\}$, $\Omega_3 := \{(0,1)^2 \times (2,3)\}$. Here, we are in the geometrical nonconforming situation. The non-mortar sides which are defined to be on Ω_2 cover only a part of the adjacent mortar sides on Ω_1 and Ω_3.

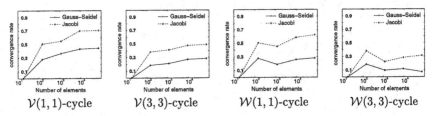

$\mathcal{V}(1,1)$-cycle $\mathcal{V}(3,3)$-cycle $\mathcal{W}(1,1)$-cycle $\mathcal{W}(3,3)$-cycle

Fig. 2.27. Jacobi and symmetric Gauß–Seidel smoother, (Example 5)

On the initial triangulation, the Lagrange multiplier space on each interface is one dimensional. We impose Dirichlet boundary values on the left and right faces of $\partial\Omega_1$ and $\partial\Omega_3$, and set $u(x,y,z) = 10$ for $(x,y,z) \in \{\partial\Omega_i, i \in \{1,3\}, x \in \{0,1\}\}$, elsewhere we impose homogeneous Neumann boundary conditions. No Dirichlet boundary conditions are given on Ω_2. Figure 2.27 shows the convergence rates for the Jacobi and the symmetric Gauß–Seidel smoother. The performance is not as good as in Example 4, but the asymptotic convergence rates seem to be independent of the refinement level. Increasing the number of smoothing steps yields considerably better results.

2.3.6 Extensions to Linear Elasticity

In this subsection, we consider the deformation of a linear elastic body as model problem. The body in its reference configuration is identified with the bounded polyhedral domain $\Omega \subset \mathbb{R}^d$, $d = 2, 3$. The planar case $d = 2$ can be interpreted as a model for an infinite long cylindric bar, $\Omega \times \mathbb{R}$. Then, the deformation of the body is described by a system of partial differential equations. The displacement field $\mathbf{u}$ of the body satisfies the following boundary value problem

$$
\begin{aligned}
-\sigma_{ij}(\mathbf{u})_{,j} &= f_i, & \text{in } \Omega\ , \\
\mathbf{u} &= 0, & \text{on } \Gamma_D\ , \\
\sigma_{ij}(\mathbf{u}) \cdot n_j &= p_i, & \text{on } \Gamma_F\ ,
\end{aligned} \tag{2.45}
$$

where $\mathbf{n}$ is the unit outer normal on the boundary of Ω. The volume force is denoted by $\mathbf{f} \in (L^2(\Omega))^d$, and $\mathbf{p} \in (L^2(\Gamma_F))^d$ is the surface traction. We denote tensor and vector quantities by bold symbols, e.g., $\boldsymbol{\tau}$ and $\mathbf{v}$, and its components by τ_{ij} and v_i, $1 \le i, j \le d$. The partial derivative with respect to x_j is abbreviated with the index $_{,j}$. Furthermore, we enforce the summation convention on all repeated indices ranging from 1 to d, and we denote by δ_{ij} the Kronecker symbol. The system (2.45) is obtained by the equation of equilibrium, the strain-displacement relation and the constitutive law. In the case of a linear elastic material, the stress tensor $\boldsymbol{\sigma}$ depends linearly on the infinitesimal strain tensor $\boldsymbol{\epsilon}(\mathbf{u}) := 1/2(\nabla \mathbf{u} + \nabla \mathbf{u}^T)$. The stress tensor $\boldsymbol{\sigma}$ is given by Hooke's law

$$\sigma_{ij}(\mathbf{u}) := E_{ijlm}\, u_{l,m} \ ,$$

where Hooke's tensor $\mathbf{E} := (E_{ijlm})_{ijlm=1}^d$, $E_{ijlm} \in L^\infty(\Omega)$, is assumed to be sufficiently smooth, symmetric and uniformly positive definite, i.e.,

$$E_{ijlm} = E_{jilm} = E_{lmij}, \quad 1 \le i, j, k, l \le d \ ,$$

and there exists a constant such that for each symmetric tensor, $\xi_{ij} = \xi_{ji}$,

$$E_{ijlm}\xi_{ij}\xi_{lm} \ge c\,\xi_{ij}\,\xi_{ij} \ .$$

In the case of a homogeneous isotropic material, Hooke's tensor has the simple form

$$E_{ijlm} = \frac{E\,\nu}{(1+\nu)(1-2\nu)}\delta_{ij}\delta_{kl} + \frac{E}{2(1+\nu)}(\delta_{ik}\delta_{jl} + \delta_{il}\delta_{jk}) \ ,$$

where $E > 0$ is Young's modulus and $\nu \in (0, 1/2)$ is the Poisson ratio. Then, the components of the stress tensor can be written in terms of E, ν and the infinitesimal strain $\boldsymbol{\epsilon}$ tensor as

$$\sigma_{ij}(\mathbf{u}) = \frac{E\,\nu}{(1+\nu)(1-2\nu)}\delta_{ij}\epsilon_{kk}(\mathbf{u}) + \frac{E}{1+\nu}\epsilon_{ij}(\mathbf{u}) \ .$$

Using the Lamé constants μ and λ, the stress tensor satisfies

$$\boldsymbol{\sigma}(\mathbf{u}) = 2\,\mu\,\boldsymbol{\epsilon}(\mathbf{u}) + \lambda\, \mathrm{tr}(\boldsymbol{\epsilon}(\mathbf{u}))\,\boldsymbol{\delta} \ ,$$

where the trace of the strain tensor is given by $\mathrm{tr}(\boldsymbol{\epsilon}(\mathbf{u})) := \epsilon_{kk}(\mathbf{u})$, and the Lamé constants are defined by

$$\mu := \frac{E}{2(1+\nu)}, \quad \lambda := \frac{E\,\nu}{(1+\nu)(1-2\nu)} \ .$$

Here, in an abuse of notation we use the same symbol for one of the Lamé constants as for the Lagrange multiplier in the mortar setting. In the following, we use Poisson's ratio and Young's modulus to specify a material.

The boundary $\partial\Omega = \overline{\Gamma}_D \cup \overline{\Gamma}_F$ is decomposed into two complementary parts, a Dirichlet part Γ_D with non zero measure and a Neumann part Γ_N, $\Gamma_N \cap \Gamma_D = \emptyset$. On Γ_D the displacement is set to zero and on Γ_N the surface traction is given. Then, the space of admissible displacements $\mathbf{H}_*^1(\Omega)$ is a subspace of $\mathbf{H}^1(\Omega) := (H^1(\Omega))^d$ and is defined by

$$\mathbf{H}_*^1(\Omega) := \left\{ \mathbf{v} \in \mathbf{H}^1(\Omega) \mid \mathbf{v}|_{\Gamma_D} = 0 \right\} .$$

Starting with (2.45), integration by parts yields the weak formulation. Let $\mathbf{u} \in \mathbf{H}_*^1(\Omega)$ be the solution of the following variational problem

$$a(\mathbf{u}, \mathbf{v}) = f(\mathbf{v}), \quad \mathbf{v} \in \mathbf{H}_*^1(\Omega) , \tag{2.46}$$

where $f(\mathbf{v}) := (\mathbf{v}, \mathbf{f})_{0;\Omega} + (\mathbf{v}, \mathbf{p})_{0;\Gamma_F}$, and the bilinear form $a(\cdot, \cdot)$ is defined by

$$a(\mathbf{w}, \mathbf{v}) := \int_\Omega E_{ijlm} w_{i,j} v_{l,m} \, dx, \quad \mathbf{w}, \mathbf{v} \in \mathbf{H}^1(\Omega) .$$

Associated with $a(\cdot, \cdot)$ is the energy norm $\|\cdot\|$, $\|\mathbf{v}\|^2 := a(\mathbf{v}, \mathbf{v})$. Applying the Lemma of Lax–Milgram, the well-posedness of (2.46) follows from the continuity of the bilinear form $a(\cdot, \cdot)$ and the linear form $f(\cdot)$ and the second Korn inequality; see, e.g., [Bra97, BS94]. We refer to [Cia88, Gur81, MH94, KO88] for a general introduction to continuum mechanics and elasticity.

The decomposition of the domain into subdomains is chosen according to the different materials or bodies. We consider two different situations. In the first situation, we assume that the different materials are glued together such that the jump of the displacements in tangential and normal direction vanishes. In Subsubsect. 2.3.6.1, the question of uniform ellipticity is addressed, and in Subsubsect. 2.3.6.2, numerical results illustrate the deformation of a body. We solve a simplified linearized form of a contact problem in Subsubsect. 2.3.6.3. In that situation, the weak solution of the continuous variational problem does not have to be in $\mathbf{H}_*^1(\Omega)$. At the interface, arbitrary displacements in tangential direction are admissible, but the jumps of the displacements in normal direction have to be zero across the interface, i.e., $[\mathbf{u} \cdot \mathbf{n}] = 0$.

For the discretization, we use the same finite element spaces of order one as before, see (1.7). The unconstrained vector valued finite element space $\mathbf{X}_h$ is defined as product space $\mathbf{X}_h := (X_h)^d$. In the case of the coupling in tangential and normal direction, the Lagrange multiplier space $\mathbf{M}_h := (M_h)^d$ is also vector valued. Then, the saddle point formulation of the mortar method can be defined in terms of the bilinear form $b(\cdot, \cdot)$

$$b(\mathbf{v}, \boldsymbol{\mu}) = \sum_{m=1}^M \langle [v_i], \mu_i \rangle_{\gamma_m}, \quad \mathbf{v} \in \mathbf{X}_h, \ \boldsymbol{\mu} \in \mathbf{M}_h .$$

Find $(\mathbf{u}_h, \boldsymbol{\lambda}_h) \in \mathbf{X}_h \times \mathbf{M}_h$ such that

$$a(\mathbf{u}_h, \mathbf{v}) + b(\mathbf{v}, \boldsymbol{\lambda}_h) = f(\mathbf{v}), \quad \mathbf{v} \in \mathbf{X}_h ,$$
$$b(\mathbf{u}_h, \boldsymbol{\mu}) \qquad\qquad = 0, \qquad \boldsymbol{\mu} \in \mathbf{M}_h . \tag{2.47}$$

Here, the bilinear form $a(\cdot, \cdot)$ is extended to the nonconforming space $\mathbf{X}_h$ by replacing the integral over Ω by its broken form $\sum_{k=1}^{K} \int_{\Omega_k}$. The second equation of the saddle point problem guarantees the weak continuity of the solution $\mathbf{u}_h$.

In analogy to Sect. 1.1, the nonconforming space $\mathbf{V}_h$ is defined as the kernel of the operator $B^T : \mathbf{X}_h \longrightarrow \mathbf{M}_h$ associated with the bilinear form $b(\cdot, \cdot)$,

$$\mathbf{V}_h := \{\mathbf{v} \in \mathbf{X}_h \mid b(\mathbf{v}, \boldsymbol{\mu}) = 0, \ \boldsymbol{\mu} \in \mathbf{M}_h\} ,$$

and we find $\mathbf{V}_h = (V_h)^d$. Under the assumption that $a(\cdot, \cdot)$ is uniformly elliptic on $\mathbf{V}_h \times \mathbf{V}_h$, i.e.,

$$a(\mathbf{v}, \mathbf{v}) \geq c \|\mathbf{v}\|_1^2 := \sum_{k=1}^{K} \|\mathbf{v}\|_{1;\Omega_k}^2, \quad \mathbf{v} \in \mathbf{V}_h ,$$

the following variational problem has a unique solution: Find $\mathbf{u}_h \in \mathbf{V}_h$ such that

$$a(\mathbf{u}_h, \mathbf{v}) = f(\mathbf{v}), \quad \mathbf{v} \in \mathbf{V}_h . \tag{2.48}$$

In the next subsubsection, we address the question of ellipticity. A uniform discrete inf-sup condition yields in combination with the ellipticity of $a(\cdot, \cdot)$ on the kernel of the operator B^T the unique solvability of (2.47). We refer to Subsect. 1.2.3 for the proof of the inf-sup condition in the scalar case. Since $\mathbf{M}_h$ and $\mathbf{X}_h$ are product spaces, the inf-sup condition follows from the scalar case. Moreover, the positive definite system (2.48) is equivalent to the saddle point problem (2.47).

2.3.6.1 Uniform Ellipticity. Here, we consider the uniform ellipticity of the bilinear form $a(\cdot, \cdot)$ on the constrained space $\mathbf{V}_h \times \mathbf{V}_h$. We start with the special case that $\partial\Omega_k \cap \Gamma_D$ has a non zero measure for all $1 \leq k \leq K$. Then, Korn's inequality can be applied to each subdomain, and we find

$$a(\mathbf{v}, \mathbf{v}) = \sum_{k=1}^{K} a_k(\mathbf{v}, \mathbf{v}) \geq C \sum_{k=1}^{K} \|\mathbf{v}\|_{1;\Omega_k}^2 = C\|\mathbf{v}\|_1^2, \quad \mathbf{v} \in \mathbf{H}_*^1(\Omega) ,$$

where $a_k(\cdot, \cdot)$ stands for the restriction of $a(\cdot, \cdot)$ to the subdomain Ω_k; see, e.g., [Bra97, BS94]. Thus $a(\cdot, \cdot)$ is elliptic on $\mathbf{H}_*^1(\Omega) \times \mathbf{H}_*^1(\Omega)$. We remark that C does not depend on the number of subdomains. Unfortunately, many interesting cases do not satisfy this assumption. However for the unique solvability of (2.48), it is sufficient to have the uniform ellipticity of $a(\cdot, \cdot)$ on $\mathbf{V}_h \times \mathbf{V}_h$.

In the scalar elliptic case, the kernel of the corresponding bilinear form is the subspace of piecewise constant functions. Its dimension is given by the number of subdomains Ω_k such that $\partial\Omega_k \cap \Gamma_D$ is empty. In the elasticity setting, the kernel is of higher dimension. The rigid body motions per subdomain define a three dimensional space in 2D and a six dimensional space in 3D. To get a better feeling for the kernel of $a(\cdot,\cdot)$, we consider the case of two unit squares Ω_1 and Ω_2 in 2D with homogeneous Dirichlet boundary condition on one side of $\partial\Omega_1 \cap \partial\Omega$ and homogeneous Neumann boundary condition elsewhere. We set $\mathbf{v}|_{\Omega_1} := 0$ and $\mathbf{v}|_{\Omega_2} := \beta(y - y_c, x_c - x)^T$, $\beta \neq 0$, where $(x_c, y_c)^T$ denotes the center of gravity of $\gamma := \partial\Omega_1 \cap \partial\Omega_2$. Then, $\mathbf{v} \in (Y)^2$ but $a(\mathbf{v}, \mathbf{v}) = 0$, and thus $a(\cdot, \cdot)$ is not elliptic on $(Y)^2 \times (Y)^2$, where the nonconforming space Y is defined by (1.6). Moreover to obtain ellipticity on $\mathbf{V}_h \times \mathbf{V}_h$, the dimension of the Lagrange multiplier space has to be larger than d. In the following, we assume that $N_m \geq 2$ in 2D, and that in 3D the triangulation at the interface is a tensor product mesh with $N_m \geq 4$. We recall that N_m is the dimension of $M_h(\gamma_m)$.

Based on this observation, we define the nonconforming space

$$\mathbf{V}_H := \left\{ \mathbf{v} \in \mathbf{H}^1_*(\Omega) \mid \int_{\gamma_m} [\mathbf{v}] \cdot \boldsymbol{\mu}\, ds = 0, \boldsymbol{\mu} \in \mathbf{M}_H(\gamma_m), 1 \leq m \leq M \right\} ,$$

where $\mathbf{M}_H(\gamma_m) := (M_H(\gamma_m))^d$ is a suitable test space. If $M_H(\gamma_m) \subset M_h(\gamma_m)$ then $\mathbf{V}_h \subset \mathbf{V}_H$, and to obtain the uniform ellipticity on $\mathbf{V}_h \times \mathbf{V}_h$ it is sufficient to show the ellipticity on $\mathbf{V}_H \times \mathbf{V}_H$. A natural choice for $M_H(\gamma_m)$ is $P_1(\gamma_m)$. Unfortunately, none of the considered Lagrange multiplier spaces satisfy $P_1(\gamma_m) \subset M_h(\gamma_m)$.

We start with the construction of a new macro Lagrange multiplier space $M_H(\gamma_m) \subset M_h(\gamma_m)$ in 2D. Let $t \in [0,1]$ be a parametrization of the 1D interface γ_m, i.e., $x \in \overline{\gamma}_m$ if and only if $x = x_0 + t_x(x_{N_m+1} - x_0)$, $t_x \in [0,1]$, where x_0 and x_{N_m+1} are the two endpoints of γ_m. Then, we decompose the set of interior vertices $\{x_i \mid 1 \leq i \leq N_m\}$ into two disjoint subsets. Introducing $\kappa_m := \max\{1 \leq i \leq N_m \mid t_{x_i} \leq 0.5\}$, we define a test function μ_H by

$$\mu_H := \left(\frac{2}{t_{x_{\kappa_m}} + t_{x_{\kappa_m+1}}} - 1 \right) \sum_{i=1}^{\kappa_m} \psi_i - \sum_{i=\kappa_m+1}^{N_m} \psi_i ,$$

where ψ_i is either the nodal standard basis function ψ_i^1 or the dual one ψ_i^3; see Subsubsect. 1.2.4.1. Figure 2.28 shows the shape of the test function μ_H. In the left, μ_H is given for the standard Lagrange multiplier space and in the right for the dual Lagrange multiplier space. It is easy to see that the mean value of μ_H is equal zero for both choices. Now, we define in terms of μ_H the macro space

$$M_H(\gamma_m) := \mathrm{span}\{\varphi_H \in \Phi_H\} ,$$

where $\Phi_H := \{1, \mu_H\}$ in the 2D case. In 3D, we work with the simplified dual basis functions given in Subsubsect. 1.2.4.2. We use the tensor product structure to define $\Phi_H := \{1, \mu_H^1, \mu_H^2, \mu_H^1 \mu_H^2\}$, where μ_H^1 and μ_H^2 are the one dimensional test functions with respect to the two different directions. Thus for the 2D and 3D case, we have $M_H(\gamma_m) \subset M_h(\gamma_m)$.

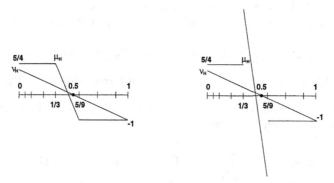

Fig. 2.28. Test function μ_H for standard (left) and dual (right) Lagrange multiplier

Lemma 2.30. *Let* $\mathbf{v} \in \mathbf{V}_H$, *and* $\mathbf{v}$ *restricted on* Ω_k, $1 \le k \le K$, *be a rigid body motion, then* $\mathbf{v} = 0$.

Proof. We start with a subdomain Ω_{k_0} such that $\partial\Omega_{k_0} \cap \Gamma_D$ has a non zero measure. Due to the Dirichlet boundary condition, $\mathbf{v}$ restricted to this subdomain is zero. On each adjacent subdomain Ω_l, we can write $\mathbf{v}|_{\Omega_l} = \mathbf{a}_l + \mathbf{w}_l$, where $\mathbf{a}_l \in \mathbb{R}^d$ and

$$\mathbf{w}_l = \alpha_l \begin{pmatrix} y - y_c \\ x_c - x \end{pmatrix}, \quad \text{and} \quad \mathbf{w}_l = \begin{pmatrix} \alpha_l^1 \\ \alpha_l^2 \\ \alpha_l^3 \end{pmatrix} \times \begin{pmatrix} x - x_c \\ y - y_c \\ z - z_c \end{pmatrix}, \quad \alpha_l, \alpha_l^i \in \mathbb{R}$$

in 2D and 3D, respectively. The center of gravity of $\gamma_{m_0} := \partial\Omega_{k_0} \cap \partial\Omega_l$ is denoted by (x_c, y_c) in 2D and by (x_c, y_c, z_c) in 3D. Then, the interface condition yields that for all adjacent subdomains, we find in 2D

$$\int_{\gamma_{m_0}} (\mathbf{a}_l + \mathbf{w}_l) \cdot \mathbf{e}_i \, d\sigma = 0, \quad 1 \le i \le d ,$$

where $\mathbf{e}_i$ denotes the i-th unit vector, and thus $\mathbf{a}_l = 0$. Introducing in 2D $\nu_H \in P_1(\gamma_{m_0})$, $\nu_H(x_0) = -\nu_H(x_{N_m+1}) = 1$, see Fig. 2.28, a straightforward computation shows that

$$\int_{\gamma_{m_0}} \mu_H \nu_H \, d\sigma = \alpha_H \, |\gamma_{m_0}| .$$

We find the lower bound $\alpha_H \geq 1/3$ if $M_h(\gamma_{m_0})$ is the piecewise linear dual Lagrange multiplier space as discussed in Subsubsect. 1.2.4.1, and $\alpha_H \geq 2/9$ if $M_h(\gamma_{m_0})$ is the standard Lagrange multiplier space and thus $\mathbf{w}_l = 0$. In 3D, $\mathbf{w}_l = 0$ follows from the 2D situation and the tensor product structure. We note that the lower bound for α_H does not depend on the mesh on γ_{m_0}. Starting from Ω_{k_0}, we can move to each other subdomain by crossing interfaces.   □

The following lemma yields the uniform ellipticity of $a(\cdot, \cdot)$ on $\mathbf{V}_h \times \mathbf{V}_h$.

Lemma 2.31. *The bilinear form $a(\cdot, \cdot)$ is uniformly elliptic on $\mathbf{V}_H \times \mathbf{V}_H$.*

Proof. The proof follows from the definition of $\mathbf{V}_H$ and the following upper bound for the broken H^1-norm

$$\|\mathbf{v}\|_1^2 \leq C_1 a(\mathbf{v}, \mathbf{v}) + C_2 \sum_{m=1}^{M} \sum_{\varphi_H \in \Phi_H} \|\int_{\gamma_m} \varphi_H[\mathbf{v}] \, d\sigma\|^2, \quad \mathbf{v} \in \mathbf{H}_*^1(\Omega) , \quad (2.49)$$

where $\| \cdot \|$ stands for the Euclidean norm in $\mathbb{R}^d$, and $0 < C_1, C_2 < \infty$ are suitable constants. The upper bound (2.49) can be established by contradiction. We assume that (2.49) is not true for any constants $0 < C_1, C_2 < \infty$. Then, there exists a sequence $\mathbf{v}^n \in \mathbf{H}_*^1(\Omega)$, $n \in \mathbb{N}$, such that $\|\mathbf{v}^n\|_1 = 1$ and

$$a(\mathbf{v}^n, \mathbf{v}^n) + \sum_{m=1}^{M} \sum_{\varphi_H \in \Phi_H} \|\int_{\gamma_m} \varphi_H[\mathbf{v}^n] \, d\sigma\|^2 < \frac{1}{n} . \quad (2.50)$$

Each $\mathbf{v}^n$ can be uniquely decomposed into a rigid body motion and a surplus. To do so, we introduce the projection operator Π_{rot} as follows: $(\Pi_{\text{rot}}\mathbf{v})_{|\Omega_k}$ is a rigid body motion on Ω_k, $1 \leq k \leq K$. If $\partial\Omega_k \cap \Gamma_D$ has a non zero measure, we set $(\Pi_{\text{rot}}\mathbf{v})_{|\Omega_k} = 0$, otherwise it satisfies

$$\int_{\Omega_k} \Pi_{\text{rot}}\mathbf{v} \, d\sigma = \int_{\Omega_k} \mathbf{v} \, d\sigma, \quad \int_{\Omega_k} \text{rot}(\Pi_{\text{rot}}\mathbf{v}) \, d\sigma = \int_{\Omega_k} \text{rot } \mathbf{v} \, d\sigma ,$$

where in 2D rot $\mathbf{v} := (-v_{1,2} + v_{2,1})$ and in 3D rot $\mathbf{v} := (v_{3,2} - v_{2,3}, v_{1,3} - v_{3,1}, v_{2,1} - v_{1,2})^T$. Then, $\mathbf{v}^n = \Pi_{\text{rot}}\mathbf{v}^n + (\mathbf{v}^n - \Pi_{\text{rot}}\mathbf{v}^n) =: \Pi_{\text{rot}}\mathbf{v}^n + \mathbf{w}^n$. Moreover by means of the second Korn inequality and the positive definiteness of Hooke's tensor, we find

$$c\|\mathbf{w}^n\|_1^2 \leq a(\mathbf{w}^n, \mathbf{w}^n) = a(\mathbf{v}^n, \mathbf{v}^n) < \frac{1}{n} .$$

Thus $\mathbf{w}^n$ converges strongly in $\mathbf{H}_*^1(\Omega)$ to zero. Using the definition of Π_{rot}, we get $\|\Pi_{\text{rot}}\mathbf{v}^n\|_1 \leq C\|\mathbf{v}^n\|_1 = C$. The range of the projection Π_{rot} is a closed finite dimensional subspace of $\mathbf{H}_*^1(\Omega)$. Its dimension is bounded by $6\,K$. Therefore, there exists a strongly convergent subsequence; the indices are denoted by n_i. Then the subsequence $\mathbf{v}^{n_i}$ converges strongly to an element $\mathbf{v} \in \mathbf{H}^1_*(\Omega)$, and the limit $\mathbf{v}$ restricted to each subdomain is a rigid body motion. Moreover, $\|\mathbf{v}\|_1 = 1$ and (2.50) yields that $\mathbf{v} \in \mathbf{V}_H$ which is a contradiction to Lemma 2.30.   □

Although the space $\mathbf{V}_H$ depends on the triangulation, the ellipticity constant can be bounded from below, due to the lower bound for α_H, independently of the triangulation.

Remark 2.32. *Unfortunately, the proof by contradiction gives only the existence of such a constant C_1, but no information if C_1 can be chosen independently of the number of subdomains. However, the decomposition into subdomains is given by physical parameters, and in many interesting applications, e.g. contact problems, the number of subdomains is fixed and small.*

It is likely that more elaborate techniques yield an ellipticity constant which is independent of the number of subdomains. For the scalar elliptic case in 2D, we refer to [Gop99] and for the three field approach to [BM00]. Both techniques are based on duality arguments and cannot be applied directly to our situation.

2.3.6.2 Numerical Results. Considering the algebraic structure of the saddle point problem (2.47) and using the same notation as in Subsect. 1.3.3, we can directly apply the previously discussed multigrid method. The following figures show some numerical results for a linear elastic body. In our first example, we consider a L-shape domain and two different materials; see Fig. 2.29. Material 1 is porcelain and its elasticity modulus and its Poisson number are given by $E = 7.9 \cdot 10^{10} N/m^2$, $\nu = 0.37$, and Material 2 is silver with $E = 5.8 \cdot 10^{10} N/m^2$, and $\nu = 0.23$. We use simplicial triangulations and piecewise linear finite elements for Material 1 and piecewise bilinear elements on rectangles for Material 2. In both cases, the trace is piecewise linear at the interfaces. Adaptive refinement techniques have been used. The refinement is controlled by a residual type error estimator for mortar finite elements; see [Woh99c]. The left picture in Fig. 2.29 illustrates the problem setting.

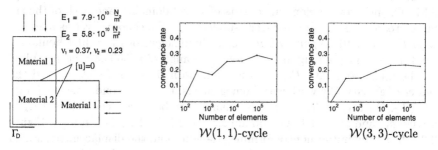

Fig. 2.29. Problem setting and convergence rates, (Example 1)

Two different surface pressures are applied at the inhomogeneous Neumann boundary part indicated by the arrows in the left of Fig. 2.29. The resulting displacements at that inhomogeneous Neumann boundary part are constant and linear, respectively; see Fig. 2.30. In both settings, the lower left corner has fixed Dirichlet boundary conditions and homogeneous Neumann boundary conditions are applied elsewhere, and the body force is zero.

Figure 2.29 illustrates the convergence rates for a $\mathcal{W}(1, 1)$-cycle with a modified ILU-type smoother and for a $\mathcal{W}(3, 3)$-cycle with a symmetric Gauß–Seidel smoother. In both cases, uniform refinement techniques have been applied. The observed convergence rates are asymptotically constant. We obtain better convergence rates in the case of three pre- and postsmoothing steps. However, the difference is not as distinct as in the scalar case for Examples 1–3; see Subsect. 2.3.5. This corresponds to the fact that we use a more robust smoother in the case of only one pre- and postsmoothing step.

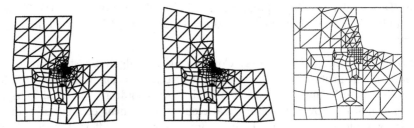

Fig. 2.30. Displacements and adaptive triangulations, (Example 1)

In Fig. 2.30, the displacements for the two different sets of boundary conditions scaled by the factor ten and the final adaptive triangulations are given. The right picture shows a zoom of the adaptive triangulation in the neighborhood of the corner singularity. Although, the triangulations at the interfaces are highly non-matching, no penetration of the two different materials occurs at the interface.

In our second example, we consider a nut-like geometry as depicted in Fig. 2.31. The non convex domain consists of 13 subdomains, and each of the six inner crosspoints has four adjacent subdomains. We choose silver as material with $E = 5.8 \cdot 10^{10} N/m^2$ and $\nu = 0.23$. Inhomogeneous Dirichlet boundary conditions corresponding to a rotation by an angle of $\pi/500$ have been applied on the inner boundary Γ_I, i.e., the outer normal on Γ_I directs toward the center of gravity. We work with homogeneous boundary conditions on $\Gamma_O := \partial\Omega \setminus \Gamma_I$. On $\Gamma_O \cap \partial\Omega_k$, we take Neumann type boundary conditions if Ω_k is a triangle, and Dirichlet type boundary conditions if Ω_k is a square. Figure 2.31 shows the initial nonconforming triangulation, the displacements scaled by the factor 100 on the final triangulation, and the multigrid convergence rates of the $\mathcal{V}$-cycle and $\mathcal{W}$-cycle with three pre- and postsmoothing steps.

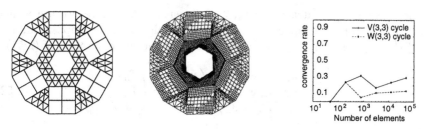

Fig. 2.31. Initial triangulation (left), distorted grid (middle) and convergence rates (right), (Example 2)

2.3.6.3 A Weaker Interface Condition. In this subsubsection, we consider a weaker coupling condition at the interface. At the interfaces, arbitrary displacements in tangential direction are admissible, but the jumps of the displacements in normal direction have to be zero. A suitable coupling condition is enforced in normal direction. This setting can be viewed as a simplified linearized form of a non linear contact problem. Figure 2.32 illustrates the modified situation at the interface. In contrast to the previous subsubsection, we have no constraints in tangential direction of the interface.

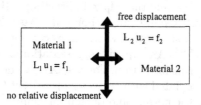

Fig. 2.32. Free tangential displacement at the interface

We define the bilinear form $b_n(\cdot, \cdot)$ corresponding to the coupling in normal direction at the interface by

$$b_n(\mathbf{v}, \mu) := \sum_{m=1}^{M} \langle [\mathbf{v} \cdot \mathbf{n}], \mu \rangle_{\gamma_m}, \qquad \mu \in M_h := \prod_{m=1}^{M} M_h(\gamma_m),$$

where $\mathbf{n}$ is the outer normal of the subdomain on the non-mortar side. Now, we replace the bilinear form $b(\cdot, \cdot)$ in (2.47) by the modified one and obtain the following saddle point problem: Find $(\mathbf{u}_h^n, \lambda_h^n) \in \mathbf{X}_h \times M_h$ such that

$$\begin{aligned}
a(\mathbf{u}_h^n, \mathbf{v}) + b_n(\mathbf{v}, \lambda_h^n) &= f(\mathbf{v}), & \mathbf{v} \in \mathbf{X}_h, \\
b_n(\mathbf{u}_h^n, \mu) &= 0, & \mu \in M_h.
\end{aligned} \tag{2.51}$$

At first glance, it has the same structure as (2.47). However, there is an essential difference. Our new bilinear form $b_n(\cdot, \cdot)$ is defined on $\mathbf{X}_h \times M_h$, where M_h is, in contrast to $\mathbf{M}_h$, a scalar space. As before, we use a dual Lagrange multiplier space. Using the same algebraic decomposition, i.e., $(\mathbf{u}_h^n)^T = ((\mathbf{u}_R^n)^T, (\mathbf{u}_N^n)^T)$, we find for the algebraic representation of (2.51)

$$
\begin{pmatrix} A_{RR} & A_{RN} & M_n \\ A_{NR} & A_{NN} & D_n \\ M_n^T & D_n^T & 0 \end{pmatrix} \begin{pmatrix} \mathbf{u}_R^n \\ \mathbf{u}_N^n \\ \lambda_h^n \end{pmatrix} = \begin{pmatrix} \mathbf{f}_R \\ \mathbf{f}_N \\ 0 \end{pmatrix} . \tag{2.52}
$$

In contrast to D in (1.46), D_n is not a diagonal matrix but a $d\,n_h \times n_h$ block diagonal matrix, where n_h is the dimension of the Lagrange multiplier space M_h. Each block is associated with an interior vertex on the non-mortar side, and the block size is given by $d \times 1$. Thus, we cannot eliminate the Lagrange multiplier as easy as in (1.40). Let $\mathcal{P} := \cup_{m=1}^M \{x_i \mid 1 \leq i \leq N_m\}$ be the set of interior vertices on the non-mortar sides. Then, we can write D_n as $D_n := \mathrm{diag}(\mathbf{d}_p)_{p \in \mathcal{P}}$, where $\mathbf{d}_p \in \mathbb{R}^d$ is defined by

$$
\mathbf{d}_p := \frac{1}{2^{d-1}} \sum_{e \in \Sigma_p} |e| \, \mathbf{n}_e .
$$

Here, Σ_p is the set of elements, i.e., edges in 2D or faces in 3D, on the non-mortar side sharing the vertex p, and $\mathbf{n}_e$ is the constant outer unit normal on the element e. We assume that $\mathbf{d}_p \neq 0$. Starting with $\mathbf{b}_1 := \mathbf{d}_p / \|\mathbf{d}_p\|$, we introduce for each vertex $p \in \mathcal{P}$ an orthonormal basis $\mathcal{B} := \{\mathbf{b}_1, \dots, \mathbf{b}_d\}$ in $\mathbb{R}^d$. The orthogonal transformation which maps $\mathcal{B}$ to the canonical basis of $\mathbb{R}^d$ is denoted by $\mathcal{O}_p \in \mathbb{R}^{d \times d}$. An explicit representation of $\mathcal{O}_p$ can be obtained, e.g., as Householder transformation. For $\mathbf{v} \in \mathbf{X}_h$, we denote by $\mathbf{v}_p \in \mathbb{R}^d$ the degrees of freedom associated with the vertex p. We set $(v_{p;n}, \mathbf{v}_{p;T}^T)^T := \mathcal{O}_p \mathbf{v}_p$ and call $v_{p;n}$ and $\mathbf{v}_{p;T}$ the normal and tangential component of $\mathbf{v}$ at the vertex p, respectively. Then, we define the global orthonormal transformation $\mathcal{O}_N$ by

$$
\mathcal{O}_N := \mathrm{diag}(\mathcal{O}_p)_{p \in \mathcal{P}} .
$$

Applying the coordinate transformation represented by $\mathrm{diag}(\mathrm{Id}, \mathcal{O}_N, \mathrm{Id})$ to (2.52), we find the symmetric system

$$
\begin{pmatrix} A_{RR} & A_{RN} \mathcal{O}_N^T & M_R \\ \mathcal{O}_N A_{NR} & \mathcal{O}_N A_{NN} \mathcal{O}_N^T & \mathcal{O}_N D_n \\ M_R^T & D_n^T \mathcal{O}_N^T & 0 \end{pmatrix} \begin{pmatrix} \mathbf{u}_R^n \\ \mathcal{O}_N \mathbf{u}_N^n \\ \lambda_h^n \end{pmatrix} = \begin{pmatrix} \mathbf{f}_R \\ \mathcal{O}_N \mathbf{f}_N \\ 0 \end{pmatrix} .
$$

The orthonormal transformation $\mathcal{O}_N$ gives a decomposition of $\mathbf{u}_N^n$ into a normal and tangential component, i.e., $\mathcal{O}_N \mathbf{u}_N^n = ((u_n^n)^T, (\mathbf{u}_T^n)^T)^T$. Due to the construction of $\mathcal{O}_N$, we have $\mathcal{O}_N^T \mathcal{O}_N = \mathrm{Id}$ and $\mathcal{O}_p \mathbf{d}_p = (\|\mathbf{d}_p\|, 0)^T$. Thus, the $d \times 1$ block matrices of $\mathcal{O}_N D_n$ are given $\|\mathbf{d}_p\| \, \mathbf{e}_1$. Observing that $D_n^T D_n$ is a diagonal matrix, the entries of which are $\|\mathbf{d}_p\|^2$, the Lagrange multiplier λ_h^n can be locally eliminated by

$$\lambda_h^n = (D_n^T D_n)^{-1} D_n^T (\mathbf{f}_N - A_{NR}\mathbf{u}_R - A_{NN}\mathbf{u}_N) \ .$$

In our last step, we can rearrange the indices. The new index group R includes now the former index group R plus the tangential components of the vectors in the former index group N, i.e., $\mathbf{u}_{R;\text{new}}^n := ((\mathbf{u}_R^n)^T, (\mathbf{u}_T^n)^T)^T$. The new index group N is a subset of the former index group N and contains the normal components, i.e., $\mathbf{u}_{N;\text{new}}^n := u_n^n$. We observe that the submatrix of $\mathcal{O}_N D_n$ corresponding to the new index group N is diagonal. Thus, we can proceed as in Section 1.3, and the proposed multigrid algorithm on the unconstrained product space can be applied. To obtain the solution from the rotated solution, we have to carry out one local postprocessing step.

Finally, we show some numerical results, illustrating the difference between the two coupling conditions at the interfaces. We start with the 2D example of Subsubsect. 2.3.6.2. Compared to Subsubsect. 2.3.6.2, we use a weaker coupling condition at the interfaces. Now, the bodies are not glued together, and free tangential displacement is permitted. The left picture in Fig. 2.33 shows the modified situation at the interface. The coupling condition in normal direction can be viewed as a kind of linearized non penetration condition of the bodies in the reference configuration. In the middle and the right picture of Fig. 2.33, the convergence rates are given for a $\mathcal{W}(1,1)$-cycle with a modified ILU-type smoother and for a $\mathcal{W}(3,3)$-cycle with a symmetric Gauß–Seidel smoother. Compared to Fig. 2.29, we observe that the stronger coupling condition at the interface results in better convergence rates. However, the difference is not significant and in both cases asymptotically constant convergence rates are obtained.

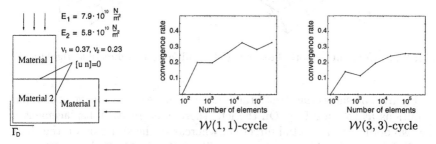

Fig. 2.33. Problem setting and convergence rates, (weak coupling)

In contrast to Fig. 2.30, penetration might be observed at the interface, see the right picture in Fig. 2.34. Although, the proposed algorithm does not solve a nonlinear contact problem, we can use the method as an inner iteration scheme within an outer scheme used to detect the actual zone of contact. Once the actual contact boundary is known, our algorithm solves the contact problem, and no penetration occurs. The drawback of this method is that in each outer iteration step a mass matrix has to be assembled.

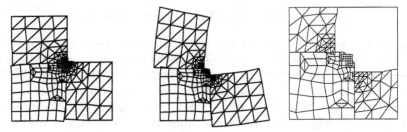

Fig. 2.34. Displacements and adaptive triangulations, (weak coupling)

A model problem in 3D is shown in Fig. 2.35. The displacement of the solution and the coarse triangulation are given for the saddle point problems (2.47) and (2.51). On the left, the Lagrange multiplier space has three degrees of freedom per node, one degree in each direction. The mortar finite element solution satisfies a weak continuity condition in tangential and normal direction. In the second situation on the right of Fig. 2.35, there is no continuity condition for the tangential displacement. Thus, we replace the bilinear form $b(\cdot,\cdot)$ in the saddle point formulation by $b_n(\cdot,\cdot)$, and work with the modifications proposed in this subsubsection.

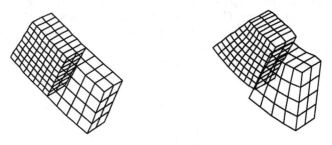

Fig. 2.35. Coupling in both directions (left) and in normal direction (right) in 3D

Figure 2.35 shows the structural difference between the two weak coupling conditions at the interface. On the left, there is no relative displacement of the two bodies in tangential direction, whereas in the situation on the right, a sliding between the two bodies is permitted. A relative displacement of the left body with respect to the right body can be observed. We remark that the nonconforming constrained space $\mathbf{V}_h$ is a subspace of $\mathbf{V}_h^n$. Here, $\mathbf{V}_h^n$ is the kernel of the operator $(B_h^n)^T : \mathbf{X}_h \longrightarrow \mathbf{M}_h$ associated with the bilinear form $b_n(\cdot,\cdot)$. In the general situation that $\partial\Omega_k \cap \Gamma_D$ is empty for some subdomain indices, the ellipticity of $a(\cdot,\cdot)$ on $\mathbf{V}_h^n \times \mathbf{V}_h^n$ is lost, and rigid body motions in tangential direction are contained in $\mathbf{V}_h^n$. We obtain unique solvability in our example by imposing Dirichlet boundary conditions on one face of each subdomain.

2.4 A Dirichlet–Neumann Type Method

In this section, we use the unsymmetric mortar formulation based on dual Lagrange multipliers given in Subsect. 1.3.1 as starting point for the construction of the iterative solver. As shown, the corresponding unsymmetric variational formulation (1.42) is closely related to a Dirichlet–Neumann coupling. One possibility to construct efficient iterative solvers for the related algebraic system (1.47) would be to apply a Dirichlet–Neumann or Neumann–Neumann type preconditioner. These techniques are well known for the standard conforming case [BGLV89, BW84, BW86, DL91, Dry88, DW95, LDV91, Le 94, QV99, SBG96, Wid88], and have been successfully adapted to non-matching triangulations and mortar finite elements in [Dry99, Dry00, LSV94].

For simplicity, we restrict ourselves to the case of two subdomains and homogeneous Dirichlet boundary conditions on $\partial\Omega$, and refer to [Dry99, Dry00] for the analysis and a generalization to many subdomains. Here, we formulate the algorithm as a block Gauß–Seidel method.

2.4.1 The Algorithm

Using the notation of Subsect. 1.3.1, the mortar finite element solution u_h solves the following block system

$$\tilde{A}u_h := \begin{pmatrix} A_D^1 & -S \\ S^T A_N^1 & A_N^2 \end{pmatrix} \begin{pmatrix} u_1 \\ u_2 \end{pmatrix} = \begin{pmatrix} \tilde{f}_1 \\ \tilde{f}_2 \end{pmatrix} =: \tilde{f} \; ; \qquad (2.53)$$

see (1.39), where the submatrices and the modified right hand side are defined by $\tilde{f}_1 := ((f_I^1)^T, 0)^T$, $\tilde{f}_2 := f_2 + S^T f_1$ and

$$S := \begin{pmatrix} 0 & 0 \\ M^T & 0 \end{pmatrix}, \quad A_N^1 := \begin{pmatrix} A_{II}^1 & A_{I\Gamma}^1 \\ A_{\Gamma I}^1 & A_{\Gamma\Gamma}^1 \end{pmatrix}, \quad \tilde{f}_2 = \begin{pmatrix} f_\Gamma^2 + M f_\Gamma^1 \\ f_I^2 \end{pmatrix},$$

A_D^1 and A_N^2 are given by (1.35) and (1.36), respectively. The scaled mass matrix M is sparse due to the use of a dual Lagrange multiplier space. We note that here M^T is a mapping from the mortar side onto the interior of the non-mortar side, $m_{pq} := \int_{\partial\Omega_1\cap\partial\Omega_2} \lambda_q \phi_p^{mor} \, d\sigma / \int_{\partial\Omega_1\cap\partial\Omega_2} \lambda_q \phi_q^{non} \, d\sigma$. Here, ϕ_p^{mor} and ϕ_q^{non} denote the nodal basis functions on the mortar and the non-mortar side, respectively. Due to the different ordering of the unknowns and the definition of the bilinear form $b(\cdot, \cdot)$, we have used a different notation in Sect. 2.3. The M^T of Sect. 2.3 has a minus sign and is extended by zero to the interior unknowns.

A damped block Gauß–Seidel method applied to (2.53) yields for $n \geq 0$

$$u_h^{n+1} = u_h^n + \begin{pmatrix} \omega_D \mathrm{Id} & 0 \\ 0 & \mathrm{Id} \end{pmatrix} \begin{pmatrix} A_D^1 & -S \\ 0 & A_N^2 \end{pmatrix}^{-1} \begin{pmatrix} \mathrm{Id} & 0 \\ 0 & \omega_N \mathrm{Id} \end{pmatrix} (\tilde{f} - \tilde{A}u_h^n) \;, \qquad (2.54)$$

where $0 < \omega_D, \omega_N \le 1$ are suitable damping parameters and u_h^0 is the start iterate. Here, we have used the upper block triangle matrix to define the iteration scheme. We remark that the matrices A_N^1 and A_N^2 correspond to Neumann boundary values on the interface, and the matrix A_D^1 is the stiffness matrix of a Dirichlet problem on Ω_1. The idea of (2.54) is illustrated in Fig. 2.36.

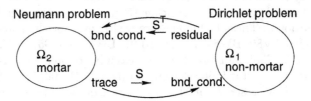

Fig. 2.36. Information transfer at the interface

Based on this observation, we can rewrite (2.54), and obtain the following Neumann–Dirichlet algorithm:

Choose damping parameters: $0 < \omega_D, \omega_N \le 1$.
Initialize: $p^1 = (1 - \omega_N)(A_N^2 u_2^0 - f_2) + \omega_N S^T(f_1 - A_N^1 u_1^0)$,
$\qquad g^0 = A_D^1 u_1^0 - \tilde{f}_1$.
For $n = 1, \dots$ *do*
$\quad$ *Solve Neumann problem on Ω_2 (mortar):*

$$A_N^2 u_2^n = f_2 + p^n \ .$$

$\quad$ *Transfer of the "Dirichlet boundary values" and damping:*

$$g^n = (1 - \omega_D)g^{n-1} + \omega_D S u_2^n \ .$$

$\quad$ *Solve Dirichlet problem on Ω_1 (non-mortar):*

$$A_D^1 u_1^n = \tilde{f}_1 + g^n \ ,$$

$\quad$ *Transfer of the "Neumann boundary values" and damping:*

$$p^{n+1} = (1 - \omega_N)p^n + \omega_N S^T(f_1 - A_N^1 u_1^n) \ .$$

We note that g^n and p^n live on Ω_1 and Ω_2, respectively. However, the start iterates can be chosen in such a way that the components of g^n and p^n associated with the interior nodes are zero. Possible choices are $u_1^0 = (A_D^1)^{-1}\tilde{f}_1$ or $u_1^0 = (A_D^1)^{-1}f_1$ and $u_2^0 = (A_N^2)^{-1}\tilde{f}_2$ or $u_2^0 = (A_N^2)^{-1}f_2$. In that case g^n and p^n can be interpreted as boundary values.

Lemma 2.33. *The Neumann–Dirichlet algorithm and the block Gauß–Seidel method (2.54) are equivalent.*

A straightforward calculation shows the assumption.

In our algorithm, the solution of a Neumann problem on Ω_2 is followed by the solution of a Dirichlet problem on Ω_1. Setting $\omega_D = 1$ gives the Neumann–Dirichlet algorithm presented in [QV99] for the conforming setting. Moreover in that case, the iterates satisfy the matching conditions at the interface, and thus are contained in the constrained space V_h. Replacing the upper block triangle matrix in (2.54) by the lower block triangle matrix yields iterates u_h^n which are in general not contained in the constrained space V_h, i.e., we start with the Dirichlet step and not with the Neumann step. The convergence analysis can be carried out within the Schwarz framework, and we refer to [Dry99] for the mortar setting. In the two subdomain case where the measure of $\partial\Omega_i \cap \partial\Omega$ is non zero for $i = 1, 2$, no coarse problem has to be solved to obtain a convergence rate in the energy norm which depends logarithmically on the ratio H/h; see [Dry99, Dry00]. Here, H stands for the maximal diameter of the subdomains and h for the minimal meshsize. Moreover, the method can serve as a preconditioner for a Schur complement system on the interface, and a conjugate gradient method can be used as acceleration.

Eliminating g^n and p^n in our algorithm yields an iterative scheme for the unknowns u_1 and u_2. In the first half step, a Neumann problem on Ω_2 has to be solved and in the second half step, we solve a Dirichlet problem on Ω_1:

$$
\begin{aligned}
u_2^n &= (1 - \omega_N)u_2^{n-1} + \omega_N(A_N^2)^{-1}(\tilde{f}_2 - S^T A_N^1 u_1^{n-1}) \\
u_1^n &= (1 - \omega_D)u_1^{n-1} + \omega_D(A_D^1)^{-1}(\tilde{f}_1 + S u_2^n)
\end{aligned}
, \quad n \geq 1 . \quad (2.55)
$$

An interesting alternative formulation can be obtained by eliminating the interior variables. Introducing a scaled Lagrange multiplier $\tilde{\lambda}_h$ defined on the non-mortar side, we can work on the interface. We recall that the mortar side is defined on Ω_2 and the non-mortar side on Ω_1. For simplicity, we restrict ourselves to the special case $\omega_D = 1$. Now, we define our preconditioned Richardson method for $\tilde{\lambda}_h$ as follows:

$$
\tilde{\lambda}_h^n := \tilde{\lambda}_h^{n-1} + \omega_N S_1(q - (S_1^{-1} + M^T S_2^{-1} M)\tilde{\lambda}_h^{n-1}) , \quad (2.56)
$$

where the Schur complement $S_i := A_{\Gamma\Gamma}^i - A_{\Gamma I}^i(A_{II}^i)^{-1}A_{I\Gamma}^i$, $i = 1, 2$, is the discrete Steklov–Poincaré operator on Ω_i, and $q := S_1^{-1}q_1 - M^T S_2^{-1}q_2$ with $q_i := f_\Gamma^i - A_{\Gamma I}^i(A_{II}^i)^{-1}f_I^i$. We note that the discrete Steklov–Poincaré operators are symmetric and positive definite. In the conforming setting, the convergence rate of the Richardson iteration (2.56) depends only on the continuity and coerciveness constants of S_1 and S_2; see [QV97]. Here, the norm of M which does not depend on the meshsize also enters. Under the assumption that the damping parameter ω_N is small enough, the convergence rate of the preconditioned Richardson iteration (2.56) is bounded independently of the meshsize, i.e.,

$$\|\tilde{\lambda}_h^n - \tilde{\lambda}_h\| \leq \rho^n \|\tilde{\lambda}_h^0 - \tilde{\lambda}_h\|$$

with $0 \leq \rho < 1$ independent of the meshsize and $\tilde{\lambda}_h := (S_1^{-1} + M^T S_2^{-1} M)^{-1} q$. The following lemma shows the relation between the iterates u_1^n, u_2^n and $\tilde{\lambda}_h^n$.

Lemma 2.34. *Under the assumptions that $\lambda_h^0 = \omega_N(0\ \mathrm{Id})(f_1 - A_N^1 u_1^0)$ and $A_N^2 u_2^0 = f_2$, the Neumann–Dirichlet algorithm with $\omega_D = 1$ and the preconditioned Richardson iteration (2.56) are equivalent. The finite element iterates can be obtained from $\tilde{\lambda}_h^n$ by*

$$\begin{aligned} u_2^n &= (A_N^2)^{-1}(f_2 + S^T(0\ (\tilde{\lambda}_h^{n-1})^T)^T) \\ u_1^n &= (A_D^1)^{-1}(\tilde{f}_1 + S u_2^n) \end{aligned}, \quad n \geq 1 .$$

Moreover, the scaled Lagrange multiplier satisfies

$$\begin{aligned} \tilde{\lambda}_h^n &= (1 - \omega_N)\tilde{\lambda}_h^{n-1} + \omega_N(0\ \mathrm{Id})(f_1 - A_N^1 u_1^n) \\ &= \omega_N(0\ \mathrm{Id}) \sum_{l=0}^{n} (1 - \omega_N)^{n-l}(f_1 - A_N^1 u_1^l) \end{aligned}, \quad n \geq 1 . \qquad (2.57)$$

The assertions can be obtained by induction and are based on (2.55), (2.56) and the definition of the Schur complements.

Comparing (2.57) with (1.40), we find that $\tilde{\lambda}_h$ is the scaled Lagrange multiplier of the mortar setting, i.e., $\tilde{\lambda}_h = D\lambda_h$ where D is a diagonal matrix, the entries of which are proportional to the local meshsize. The assumption $A_N^2 u_2^0 = f_2$ in Lemma 2.34 can be weakened. For the equivalence, its is sufficient if there exists a y_1 such that $A_N^2 u_2^0 = f_2 + S^T y_1$. A three term recursion for $\tilde{\lambda}_h^n$ can be obtained for more general damping parameters.

Remark 2.35. *This type of algorithm can be used to construct an efficient iterative solver for nonlinear problems. It can be applied to solve multi body contact problems. Then in each iteration step, a linear Neumann problem and a Signorini problem have to be solved. The Neumann data on the mortar side are obtained in terms of S^T and the Lagrange multiplier on the non-mortar side. Applying S to the Dirichlet values on the mortar side gives the one-sided obstacle for the Signorini problem on the non-mortar side. We refer to [KW00c] for the nonlinear Dirichlet–Neumann algorithm and numerical results illustrating the deformation of the bodies and the boundary stresses at the contact zone. An extension of the algorithm to 3D and Coulomb friction can be found in [KW01]. The nonlinear Signorini problem can be solved efficiently by monotone multigrid methods; see [Kor97, KK99, KK00, Kra01].*

2.4.2 Numerical Results

In this subsection, we present numerical results for the proposed Neumann–Dirichlet algorithm. We consider the extremely simple case of two unit squares Ω_i, $i = 1, 2$, and $-\Delta u = f$ on Ω, $\overline{\Omega} := \overline{\Omega}_1 \cup \overline{\Omega}_2$. The right hand side f is chosen to be constant on each subdomain, $f_{|\Omega_1} = -1$ and $f_{|\Omega_2} = 1$. Inhomogeneous Dirichlet boundary conditions are imposed on the sides of the unit squares parallel to the interface, and homogeneous Neumann boundary condition are applied elsewhere. We work with non-matching triangulations at the interface and use uniform refinement techniques. The coarse triangulation has 4 elements on the non-mortar subdomain Ω_1 and 9 elements on the mortar subdomain Ω_2. In the Neumann–Dirichlet algorithm, the start iterate u_1^0 is given by a random vector on each level, the values of the components are contained in $[-5, 5]$. Furthermore, we set $p^1 = S^T(f_1 - A_N^1 u_1^0)$ which corresponds to the choice $u_2^0 = (A_N^2)^{-1}(f_2 + S^T(f_1 - A_N^1 u_1^0))$. Replacing the upper block triangle matrix in (2.54) by the lower block triangle matrix yields a Dirichlet–Neumann algorithm. In that case, we have to initialize g^1. We define $g^1 = Su_2^0$, where u_2^0 is a random vector. This choice corresponds to $u_1^0 = (A_D^1)^{-1}(\tilde{f}_1 + Su_2^0)$.

In the Dirichlet–Neumann algorithm, we start with a Dirichlet problem on the non-mortar subdomain Ω_1 followed by a Neumann problem on the mortar subdomain Ω_2. We compare the proposed Neumann–Dirichlet algorithm, $\omega_N \in \{0.7, 0.75, 0.8, 0.85, 0.9\}$ and $\omega_D = 1$, with the Dirichlet–Neumann algorithm, $\omega_D \in \{0.7, 0.75, 0.8, 0.85, 0.9\}$ and $\omega_N = 1$. In Tables 2.5 and 2.6, the number of required iteration steps to obtain an error reduction of 10^{-6} is given for different damping parameters. The numbers in parenthesis show the numerical results for a different coarse triangulation. In that case, we start with 25 elements on Ω_1 and 4 elements on Ω_2.

Table 2.5. Number of iteration steps, (trace norm)

level	Dirichlet–Neumann ω_D					Neumann–Dirichlet ω_N				
	0.7	0.75	0.8	0.85	0.9	0.7	0.75	0.8	0.85	0.9
1	9 (10)	7 (9)	6 (7)	6 (6)	8 (6)	9 (10)	7 (8)	6 (7)	6 (5)	7 (7)
2	9 (11)	7 (9)	5 (8)	6 (6)	8 (7)	9 (9)	7 (8)	5 (7)	6 (6)	8 (7)
3	9 (10)	7 (9)	5 (7)	6 (6)	8 (7)	9 (9)	7 (8)	6 (7)	5 (6)	7 (7)
4	9 (10)	7 (9)	6 (7)	6 (6)	8 (7)	9 (9)	7 (8)	6 (7)	5 (6)	7 (7)
5	9 (11)	7 (9)	5 (8)	6 (6)	8 (7)	9 (10)	7 (8)	6 (7)	5 (6)	7 (7)
6	9 (10)	7 (9)	6 (8)	6 (6)	8 (7)	9 (9)	7 (8)	6 (7)	6 (5)	7 (7)
7	9 (10)	7 (9)	6 (7)	6 (6)	8 (7)	9 (9)	7 (8)	6 (7)	5 (6)	7 (7)

In Table 2.5, the error is measured in a weighted L^2-norm for the trace. The weighting factor is given by $1/h_2$ where h_2 is the local meshsize on the mortar side, i.e.,

$$\sum_{e \in \mathcal{S}_{h_2}} \frac{1}{h_e} \|u_2 - u_2^n\|_{0;e}^2 \leq 10^{-12} \sum_{e \in \mathcal{S}_{h_2}} \frac{1}{h_e} \|u_2 - u_2^1\|_{0;e}^2 . \tag{2.58}$$

Here, $\mathcal{S}_{h_2}$ is the set of edges at the interface on the mortar side, and u_2 is the exact discrete solution on the mortar side. We remark that this weighted

L^2-norm is equivalent to the Euclidean vector norm of $u_2 - u_2^n$ restricted to the interface nodes on the mortar side.

Table 2.6. Number of iteration steps, (Lagrange multiplier norm)

level	Dirichlet–Neumann ω_D					Neumann–Dirichlet ω_N				
	0.7	0.75	0.8	0.85	0.9	0.7	0.75	0.8	0.85	0.9
1	9 (11)	7 (9)	5 (8)	6 (6)	8 (6)	9 (10)	7 (9)	5 (7)	6 (6)	8 (7)
2	9 (11)	7 (9)	5 (8)	6 (7)	8 (6)	9 (10)	7 (9)	5 (8)	6 (6)	8 (7)
3	8 (10)	7 (9)	5 (7)	6 (6)	8 (6)	9 (11)	7 (9)	6 (8)	6 (6)	8 (7)
4	8 (11)	7 (9)	5 (8)	6 (6)	8 (6)	9 (10)	7 (9)	5 (8)	6 (6)	8 (7)
5	8 (11)	7 (9)	5 (8)	6 (6)	8 (6)	9 (11)	7 (9)	5 (8)	6 (6)	8 (7)
6	8 (11)	7 (9)	5 (8)	6 (6)	8 (6)	9 (10)	7 (9)	5 (8)	6 (6)	8 (6)
7	8 (11)	7 (9)	5 (8)	6 (6)	8 (6)	9 (11)	7 (8)	5 (8)	6 (6)	8 (7)

In Table 2.6, the error in the Lagrange multiplier defines the stopping criteria for the iterative solver. It is measured in a weighted L^2-norm. The weighting factor is given by the local meshsize on the non-mortar side. We note that the Lagrange multiplier is an approximation of the flux. It can be obtained from the residuum of a Neumann problem on the non-mortar side by a diagonal scaling, i.e., $\lambda_h = D^{-1}(0\,\mathrm{Id})(f_1 - A_N^1 u_1)$, where u_1 is the exact discrete solution on the non-mortar subdomain Ω_1. Moreover if we set $\lambda_h^n = D^{-1}(0\,\mathrm{Id})(f_1 - A_N^1 u_1^n)$, we have the following norm equivalence

$$c\,\|(A_N^1(u_1 - u_1^n))_\Gamma\|^2 \le \sum_{e \in \mathcal{S}_{h_1}} h_e \|\lambda_h - \lambda_h^n\|_{0;e}^2 \le C\,\|(A_N^1(u_1 - u_1^n))_\Gamma\|^2 .$$

Here, $\mathcal{S}_{h_1}$ is the set of edges at the interface on the non-mortar side. This equivalence motivates the stopping criteria

$$\|A_N^1(u_1 - u_1^n)\| \le 10^{-6}\|A_N^1(u_1 - u_1^1)\| . \tag{2.59}$$

We note that the choice of our start vectors guarantee that $A_N^1(u_1 - u_1^n) = 0$ for all interior nodes on Ω_1, i.e., $\|(A_N^1(u_1 - u_1^n))_\Gamma\| = \|A_N^1(u_1 - u_1^n)\|$.

The number of required iteration steps is independent of the refinement level for all damping parameters. But it depends highly on the damping parameter, see also Figs. 2.37 and 2.38. Both algorithms, Dirichlet–Neumann and Neumann–Dirichlet, require approximately the same number of iteration steps to obtain the required error reduction. The choice of the norm for the error plays only a minor role, and almost the same results are obtained for the stopping criterias (2.58) and (2.59).

Figures 2.37 and 2.38 show the computed errors and the best least-square fit for different damping parameters on level 5 and level 7. The computed errors are marked by different symbols whereas the corresponding best fit is given by a solid line. In Fig. 2.37, the error reduction is measured in the

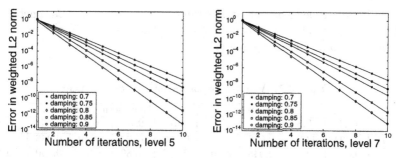

Fig. 2.37. Error reduction (trace norm) on level 5 (left) and level 7 (right) for the Dirichlet–Neumann algorithm

specified L^2-norm for the trace on the mortar side. The results for the error in the Lagrange multiplier on the non-mortar side are given in Fig. 2.38.

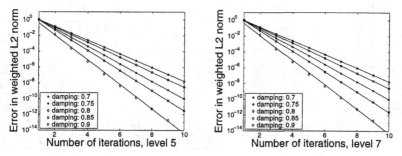

Fig. 2.38. Error reduction (Lagrange multiplier norm) on level 5 (left) and level 7 (right) for the Dirichlet–Neumann algorithm

Finally, Tables 2.7 and 2.8 give the asymptotic convergence rates. The lines of the two tables correspond to the levels $1-7$. To define the convergence rate q, we use a first order logarithmic polynomial least-square fit based on 10 iteration steps. Then, $q := 10^a$ where

$$\sum_{k=1}^{10}(e_k - a\,k - b)^2 = \min_{\alpha,\beta\in\mathbb{R}}\sum_{k=1}^{10}(e_k - \alpha\,k - \beta)^2\ ,$$

and e_k is the logarithm of the iteration error in one of the specified norms. Table 2.7 shows the convergence rates for different damping factors in the trace norm on the mortar side, and Table 2.8 gives the corresponding results for the Lagrange multiplier norm on the non-mortar side. The convergence rates depend highly on the choice of the damping parameter. In practice, the choice of a good damping parameter is a delicate point. The influence of the damping parameter can be considerably weakened by using the block Gauß–Seidel method as a preconditioner for a Krylov subspace method. Another possibility might be to use adaptive strategies for the damping parameter. We note that in general u_1 and u_2 are unknown. Then, $u_2^n - u_2^{n-1}$ and $u_1^n - u_1^{n-1}$ can be used to define a stopping criteria.

Table 2.7. Asymptotic convergence rates q, (trace)

Dirichlet–Neumann ω_D					Neumann–Dirichlet ω_N				
0.7	0.75	0.8	0.85	0.9	0.7	0.75	0.8	0.85	0.9
0.151	0.094	0.038	0.052	0.109	0.160	0.100	0.040	0.046	0.101
0.142	0.084	0.034	0.048	0.108	0.150	0.089	0.028	0.050	0.107
0.145	0.088	0.036	0.051	0.109	0.159	0.099	0.039	0.042	0.087
0.147	0.091	0.037	0.053	0.113	0.160	0.100	0.040	0.044	0.090
0.146	0.089	0.036	0.053	0.113	0.160	0.100	0.040	0.040	0.084
0.149	0.092	0.037	0.051	0.109	0.158	0.099	0.039	0.047	0.100
0.148	0.091	0.037	0.052	0.111	0.160	0.100	0.040	0.042	0.088

Table 2.8. Asymptotic convergence rates q, (Lagrange multiplier norm)

Dirichlet–Neumann ω_D					Neumann–Dirichlet ω_N				
0.7	0.75	0.8	0.85	0.9	0.7	0.75	0.8	0.85	0.9
0.139	0.081	0.035	0.049	0.109	0.146	0.088	0.036	0.045	0.105
0.140	0.079	0.026	0.046	0.106	0.142	0.083	0.027	0.050	0.110
0.139	0.078	0.027	0.050	0.110	0.154	0.095	0.037	0.050	0.106
0.138	0.078	0.028	0.051	0.111	0.141	0.085	0.035	0.051	0.112
0.138	0.078	0.028	0.053	0.113	0.145	0.089	0.036	0.053	0.113
0.138	0.078	0.027	0.051	0.111	0.140	0.081	0.032	0.050	0.111
0.138	0.078	0.028	0.052	0.112	0.142	0.085	0.034	0.052	0.111

Remark 2.36. *We remark that the application of the operators S and S^T is extremely cheap. This is due to the use of dual basis functions for the Lagrange multiplier.*

2.5 A Multigrid Method for the Mortar Saddle Point Formulation

In Sect. 2.3 and in Sect. 2.4, two different iterative solvers for mortar discretization techniques have been discussed. We have considered the case of dual Lagrange multipliers. The theoretical results also hold if standard Lagrange multipliers are used. But in that case, the computational cost for one iteration step is considerably higher. This is due to the fact, that the scaling of the mass matrix on the mortar side involves the inverse of a mass matrix on the non-mortar side. In this section, we analyze a multigrid method based on the mortar saddle point formulation with standard Lagrange multiplier spaces. In contrast to the two previous sections, the scaled mass matrix does not enter in the computation.

There are different approaches to the efficient solution of the indefinite saddle point problem (1.9). One possibility is to use a good preconditioner for the exact Schur complement as analyzed in [Kuz95a, Kuz95b] and further exploited in [EHI+98, EHI+00, HIK+98]. A different technique is based on an idea of Braess and Sarazin which is presented for the Stokes problem in [BS97]. It has been successfully adapted to mortar situations in [BD98, BDW99, WW98]. In that approach, a modified Schur complement system has to be solved exactly in each smoothing step. More recently, a simplified version of this idea has been studied in [WW99]. The modification is based on ideas introduced in [Zul01], and a more general approximation property analyzed in [Woh00b].

Here, we study a multigrid method for the saddle point problem (1.9). Two different types of smoothers are discussed; a block diagonal one and one reflecting the saddle point structure. There is no need to solve a Schur complement system exactly. The first smoother works on the squared system. In the second case, the exact solution of the modified Schur complement system is replaced by an iteration, resulting in an inner and outer iteration cycle. This multigrid method is given for the standard mortar formulation as presented in Sect. 1.1. In contrast to [BD98, BDW99, WW98], we are not working in the subspace on which the saddle point problem is positive definite; the iterates do not have to satisfy the constraints at the interfaces, and we do not have to solve a Schur complement system exactly. To obtain convergence results, it is therefore necessary to establish an appropriate approximation property for both variables u_h and λ_h. This issue was originally addressed in [Woh00b]; a weaker approximation property has also been studied in [BD98, BDW99]. For both smoothers, we obtain level independent convergence rates for the $\mathcal{W}$-cycle provided that the number of smoothing steps is large enough. We assume full H^2-regularity. In this section, the multigrid analysis is carried out in the operator setting.

We work with the standard Lagrange multiplier spaces M_l, $0 \leq l \leq j$, as defined by (1.2), and the unconstrained product spaces X_l which are associated with a nested family of quasi-uniform triangulations $\mathcal{T}_l$, $h_l = 2h_{l+1}$. The triangulations on the non-mortar sides are called $\mathcal{S}_{m;l}$. We then find for the finite element spaces that

$$X_0 \times M_0 \subset X_1 \times M_1 \subset \cdots \subset X_l \times M_l \subset \cdots \subset X_j \times M_j \ .$$

The spaces $X_l \times M_l$ are Hilbert spaces equipped with the mesh dependent bilinear forms

$$((v, \mu), (w, \nu))_{h_l; \Omega \times \mathcal{S}} := (v, w)_0 + (\mu, \nu)_{h_l^{-\frac{3}{2}}; \mathcal{S}}, \quad (v, \mu), (w, \nu) \in X_l \times M_l \ ,$$

where the mesh dependent bilinear form $(\cdot, \cdot)_{h_l^{-3/2}; \mathcal{S}}$ on $M_l \times M_l$ is given by

$$(\mu, \nu)_{h_l^{-\frac{3}{2}};\mathcal{S}} := \sum_{m=1}^{M} \sum_{e \in \mathcal{S}_{m;l}} h_e^3 \, (\mu, \nu)_{0;e}, \quad \mu, \nu \in M_l \ .$$

The corresponding norms are denoted by $\| \cdot \|_{h_l;\Omega \times \mathcal{S}}$ and $\| \cdot \|_{h_l^{-3/2};\mathcal{S}}$.

Before focusing on the approximation property, we consider suitable transfer operators. Let $I_{l-1}^l : X_{l-1} \longrightarrow X_l$ and $J_{l-1}^l : M_{l-1} \longrightarrow M_l$ be the natural injections. We then define I_l^{l-1} and J_l^{l-1} by

$$(I_l^{l-1} v_l, w_{l-1})_0 = (v_l, I_{l-1}^l w_{l-1})_0, \qquad w_{l-1} \in X_{l-1} \ ,$$
$$(J_l^{l-1} \mu_l, \nu_{l-1})_{h_{l-1}^{-\frac{3}{2}};\mathcal{S}} = (\mu_l, J_{l-1}^l \nu_{l-1})_{h_l^{-\frac{3}{2}};\mathcal{S}}, \qquad \nu_{l-1} \in M_{l-1} \ .$$

Since the spaces are nested, we have $w_{l-1} = I_{l-1}^l w_{l-1}$, $w_{l-1} \in X_{l-1}$, and $\nu_{l-1} = J_{l-1}^l \nu_{l-1}$, $\nu_{l-1} \in M_{l-1}$. A straightforward calculation now shows that if $(\widehat{w}_l, \widehat{\nu}_l) \in X_l \times M_l$ satisfies the saddle point problem

$$a(\widehat{w}_l, v_l) + b(v_l, \widehat{\nu}_l) = (\widehat{d}_l, v_l)_0, \qquad v_l \in X_l \ ,$$
$$b(\widehat{w}_l, \mu_l) \qquad\qquad = (\widehat{\delta}_l, \mu_l)_{h_l^{-\frac{3}{2}};\mathcal{S}}, \qquad \mu_l \in M_l \ ,$$

and if $(\widehat{w}_{l-1}, \widehat{\nu}_{l-1}) \in X_{l-1} \times M_{l-1}$ satisfies

$$a(\widehat{w}_{l-1}, v_{l-1}) + b(v_{l-1}, \widehat{\nu}_{l-1}) = (I_l^{l-1} \widehat{d}_l, v_{l-1})_0, \qquad v_{l-1} \in X_{l-1} \ ,$$
$$b(\widehat{w}_{l-1}, \mu_{l-1}) \qquad\qquad = (J_l^{l-1} \widehat{\delta}_l, \mu_{l-1})_{h_{l-1}^{-\frac{3}{2}};\mathcal{S}}, \qquad \mu_{l-1} \in M_{l-1} \ ,$$

then $(w_l, \nu_l) \in X_l \times M_l$, $w_l := \widehat{w}_l - I_{l-1}^l \widehat{w}_{l-1}$, $\nu_l := \widehat{\nu}_l - J_{l-1}^l \widehat{\nu}_{l-1}$ satisfies

$$a(w_l, v_l) + b(v_l, \nu_l) = (d_l, v_l)_0, \qquad v_l \in X_l \ ,$$
$$b(w_l, \mu_l) \qquad\qquad = (\delta_l, \mu_l)_{h_l^{-\frac{3}{2}};\mathcal{S}}, \qquad \mu_l \in M_l \ ,$$

with $((d_l, \delta_l), (v_{l-1}, \mu_{l-1}))_{h_l;\Omega \times \mathcal{S}} = 0$, for all $(v_{l-1}, \mu_{l-1}) \in X_{l-1} \times M_{l-1}$.

After these preliminary remarks, we will establish a general approximation property and introduce two different kinds of smoothing operators.

2.5.1 An Approximation Property

The following lemma can be found in an abstract form, as well as in the special case of mixed finite element approximations of the Stoke's equation, in [Ver84, Lemma 4.2]. Here, we adapt it to mortar finite elements. It shows that the coarse grid correction in the multigrid framework yields a good approximation of the solution of the defect equation on the fine level.

Lemma 2.37. (Approximation property)
Let $(d_l, \delta_l) \in X_l \times M_l$ be orthogonal to $X_{l-1} \times M_{l-1}$, i.e.,

$$((d_l, \delta_l), (v_{l-1}, \mu_{l-1}))_{h_l; \Omega \times S} = 0, \qquad (v_{l-1}, \mu_{l-1}) \in X_{l-1} \times M_{l-1} \ ,$$

and let $(w_l, \nu_l) \in X_l \times M_l$ be the solution of: Find $(w_l, \nu_l) \in X_l \times M_l$ such that

$$\begin{aligned}
a(w_l, v_l) + b(v_l, \nu_l) &= (d_l, v_l)_0, & v_l &\in X_l \ , \\
b(w_l, \mu_l) &= (\delta_l, \mu_l)_{h_l^{-\frac{3}{2}}; S}, & \mu_l &\in M_l \ .
\end{aligned} \qquad (2.60)$$

Then, there exists a constant C satisfying

$$\|(w_l, \nu_l)\|_{h_l; \Omega \times S} \le C \, h_l^2 \, \|(d_l, \delta_l)\|_{h_l; \Omega \times S} \ .$$

Proof. The quasi-uniformity of the triangulations yields

$$c \left(\|w_l\|_0 + h_l^{\frac{3}{2}} \|\nu_l\|_{0; S} \right) \le \|(w_l, \nu_l)\|_{h_l; \Omega \times S} \le C \left(\|w_l\|_0 + h_l^{\frac{3}{2}} \|\nu_l\|_{0; S} \right) \ .$$

The estimate for the Lagrange multiplier term will be based on the discrete inf-sup condition (1.10) whereas the bound for $\|w_l\|_0$ is based on duality techniques. We start with an estimate for the upper bound of $h_l^{3/2} \|\nu_l\|_{0; S}$. The continuity of $a(\cdot, \cdot)$, the discrete inf-sup condition (1.10), and the orthogonality of d_l on X_{l-1}, i.e., $(d_l, v_{l-1})_0 = 0$, $v_{l-1} \in X_{l-1}$, yield an upper bound for $\|\nu_l\|_{h_l^{-1/2}; S}$ in terms of $\|w_l\|_1$ and $\|d_l\|_0$:

$$\begin{aligned}
\|\nu_l\|_{h_l^{-\frac{1}{2}}; S} &\le C \sup_{\substack{v_l \in X_l \\ v_l \ne 0}} \frac{b(v_l, \nu_l)}{\|v_l\|_1} = C \inf_{v_{l-1} \in X_{l-1}} \sup_{\substack{v_l \in X_l \\ v_l \ne 0}} \frac{(d_l, v_l - v_{l-1})_0 - a(w_l, v_l)}{\|v_l\|_1} \\
&\le C \left(\|w_l\|_1 + h_l \|d_l\|_0 \right) \ .
\end{aligned} \qquad (2.61)$$

The second equation in (2.60) yields $b(w_l, \mu_{l-1}) = 0$, $\mu_{l-1} \in M_{l-1}$, and thus $w_{l-1} \in Y$. Recalling that the bilinear form $a(\cdot, \cdot)$ is elliptic on $Y \times Y$, see (1.5), and using (2.61), we find for $w_{l-1} \in X_{l-1}$

$$\begin{aligned}
c \, \|w_l\|_1^2 &\le a(w_l, w_l) = (d_l, w_l)_0 - b(w_l, \nu_l) = (d_l, w_l - w_{l-1})_0 - (\delta_l, \nu_l)_{h_l^{-\frac{3}{2}}; S} \\
&\le C h_l \left(\|d_l\|_0 + h_l^{\frac{3}{2}} \|\delta_l\|_{0; S} \right) \|w_l\|_1 + C h_l^{\frac{7}{2}} \|d_l\|_0 \|\delta_l\|_{0; S} \ .
\end{aligned}$$

To obtain an upper bound for $\|w_l\|_1$ in terms of $h_l^{3/2} \|\delta_l\|_{0; S}$ and $\|d_l\|_0$, it is sufficient to consider the quadratic polynomial

$$g(s) := s^2 - \alpha(a_1 + a_2)s - \alpha a_1 a_2, \qquad s \in \mathbb{R} \ ,$$

where $a_1 := h_l^{5/2} \|\delta_l\|_{0; S} \ge 0$, $a_2 := h_l \|d_l\|_0 \ge 0$, and α is a positive constant. In the case that $a_1 = a_2 = 0$, the unique solvability of (2.60) gives $\|w_l\|_1 = 0$. For $a_1 + a_2 > 0$, an easy calculation shows that $g(s) > 0$ for $s \ge (1 + \alpha)(a_1 + a_2)$. Thus, the following upper bound for $\|w_l\|_1$ holds in terms of $\|\delta_l\|_{0; S}$ and $\|d_l\|_0$:

$$\|w_l\|_1 \leq Ch_l \left(\|d_l\|_0 + h_l^{\frac{3}{2}} \|\delta_l\|_{0;\mathcal{S}} \right) \ . \tag{2.62}$$

Combining (2.62) with the upper bound (2.61) for $\|\nu_l\|_{h_l^{-1/2};\mathcal{S}}$, we find

$$h_l^{\frac{3}{2}} \|\nu_l\|_{0;\mathcal{S}} \leq C \, h_l^2 \left(\|d_l\|_0 + h_l^{\frac{3}{2}} \|\delta_l\|_{0;\mathcal{S}} \right) \ . \tag{2.63}$$

In our next step, we focus on an estimate for $\|w_l\|_0$. To obtain an upper bound for $\|w_l\|_0$, we use Aubin–Nitsche type arguments. Let $\hat{w} \in H_0^1(\Omega)$ be the solution of the continuous variational problem: Find $\hat{w} \in H_0^1(\Omega)$ such that

$$a(\hat{w}, v) = (w_l, v)_0, \quad v \in H_0^1(\Omega) \ .$$

Taking into account that, in general, $w_l \in X_l$ is not contained in $H_0^1(\Omega)$, we get

$$\|w_l\|_0^2 = a(\hat{w}, w_l) + b(w_l, \hat{\nu}) \ ,$$

where $\hat{\nu} := a\nabla\hat{w} \cdot \mathbf{n}$ is the flux of $\hat{w}$ across the interfaces. Using that $a(w_l, v_{l-1}) + b(v_{l-1}, \nu_l) = 0$, $v_{l-1} \in X_{l-1}$, $b(w_l, \mu_{l-1}) = 0$, $\mu_{l-1} \in M_{l-1}$, and observing that $b(\hat{w}, \nu_l) = 0$, $\nu_l \in M_l$, we find for $v_{l-1} \in X_{l-1}$ and $\mu_{l-1} \in M_{l-1}$ that

$$\begin{aligned}
\|w_l\|_0^2 &= a(\hat{w} - v_{l-1}, w_l) + b(w_l, \hat{\nu} - \mu_{l-1}) + b(\hat{w} - v_{l-1}, \nu_l) \\
&\leq C \left(\|\hat{w} - v_{l-1}\|_1 \|w_l\|_1 + \|[w_l]\|_{0;\mathcal{S}} \|\hat{\nu} - \mu_{l-1}\|_{0;\mathcal{S}} \right. \\
&\quad \left. + \|[\hat{w} - v_{l-1}]\|_{0;\mathcal{S}} \|\nu_l\|_{0;\mathcal{S}} \right) \ .
\end{aligned} \tag{2.64}$$

Here, we have used that the spaces are nested. We choose $v_{l-1} \in X_{l-1}$ as a local quasi-projection of $\hat{w}$ such that

$$\begin{aligned}
\|\hat{w} - v_{l-1}\|_1 &\leq Ch_{l-1}\|\hat{w}\|_2 \ , \\
\|[\hat{w} - v_{l-1}]\|_{0;\mathcal{S}} &\leq Ch_{l-1}^{\frac{3}{2}}\|\hat{w}\|_2 \ ;
\end{aligned} \tag{2.65}$$

see, e.g., [SZ90], and $\mu_{l-1} \in M_{l-1}$ satisfies

$$\|\hat{\nu} - \mu_{l-1}\|_{0;\mathcal{S}} \leq Ch_{l-1}^{\frac{1}{2}}\|\hat{\nu}\|_{\frac{1}{2};\mathcal{S}} \leq Ch_{l-1}^{\frac{1}{2}}\|\hat{w}\|_2 \ . \tag{2.66}$$

Remark 1.13 shows that M_{l-1} is also a suitable Lagrange multiplier space for X_l. Thus, we can apply Lemma 1.7, for $\widetilde{V}_l := \{v \in X_l, \ b(v, \mu) = 0, \ \mu \in M_{l-1}\}$, and obtain

$$\|[w_l]\|_{0;\mathcal{S}} \leq Ch_l^{\frac{1}{2}}\|w_l\|_1 \ .$$

In a last step, we use the H^2-regularity, $\|\hat{w}\|_2 \leq C\|w_l\|$, and a trace theorem. Combing the upper bounds (2.64)–(2.66) yields

$$\|w_l\|_0^2 \leq Ch_l \left(\|w_l\|_1 + \|\nu_l\|_{h_l^{-\frac{1}{2}};\mathcal{S}} \right) \|w_l\|_0 \ ,$$

which, together with (2.62) and (2.63), proves the assertion.   □

Before introducing our smoothing operators, we consider the operator K_l associated with the saddle point problem (2.60) on level l: Let $A_l : X_l \longrightarrow X_l$, $B_l : M_l \longrightarrow X_l$, $B_l^* : X_l \longrightarrow M_l$ be the operators defined by

$$(A_l v_l, w_l)_0 := a(v_l, w_l) \ ,$$
$$(B_l \mu_l, w_l)_0 := b(w_l, \mu_l) \ , \quad (B_l^* w_l, \mu_l)_{h_l^{-\frac{3}{2}};S} := b(w_l, \mu_l) \ .$$

Then, the self-adjoint non-singular operator $K_l : X_l \times M_l \longrightarrow X_l \times M_l$, associated with the saddle point problem (2.60), is given by

$$K_l(v_l, \mu_l) := (A_l v_l + B_l \mu_l, B_l^* v_l), \quad (v_l, \mu_l) \in X_l \times M_l \ . \tag{2.67}$$

The solution (w_l, ν_l) of the saddle point problem (2.60) satisfies

$$K_l(w_l, \nu_l) = (d_l, \delta_l) \ ,$$

and thus $\|K_l^{-1}(d_l, \delta_l)\|_{h_l;\Omega \times S} \leq C h_l^2 \|(d_l, \delta_l)\|_{h_l;\Omega \times S}$ for those $(d_l, \delta_l) \in X_l \times M_l$ which are orthogonal onto $X_{l-1} \times M_{l-1}$ with respect to $(\cdot, \cdot)_{h_l;\Omega \times S}$. Equivalently, we find

$$\|(w_l, \nu_l)\|_{h_l;\Omega \times S} \leq C h_l^2 \|K_l(w_l, \nu_l)\|_{h_l;\Omega \times S} \ , \tag{2.68}$$

for $(w_l, \nu_l) \in X_l \times M_l$ satisfying $(K_l(w_l, \nu_l), (v_{l-1}, \mu_{l-1}))_{h_l;\Omega \times S} = 0$, for all $(v_{l-1}, \mu_{l-1}) \in X_{l-1} \times M_{l-1}$.

2.5.2 Smoothing and Stability Properties

The second basic tool to establish convergence within the multigrid framework is the smoothing property. In this subsection, we introduce two types of smoothers. We show that they are stable and satisfy a suitable smoothing property. Before we give the definition of the smoothers, we consider the operator B_l in more detail. The stability of the block diagonal smoother can be easily established by means of the properties of the operator B_l. The following lemma shows that the condition number of B_l is uniformly bounded.

Lemma 2.38. *There exist constants such that*

$$c \frac{1}{h_l^2} \|\mu_l\|_{h_l^{-\frac{3}{2}};S} \leq \|B_l \mu_l\|_0 \leq C \frac{1}{h_l^2} \|\mu_l\|_{h_l^{-\frac{3}{2}};S} \ . \tag{2.69}$$

Proof. The upper bound is obtained by using an inverse estimate and the definition of B_l:

$$\|B_l \mu_l\|_0 = \sup_{\substack{w_l \in X_l \\ w_l \neq 0}} \frac{(B_l \mu_l, w_l)_0}{\|w_l\|_0} = \sup_{\substack{w_l \in X_l \\ w_l \neq 0}} \frac{b(w_l, \mu_l)}{\|w_l\|_0}$$
$$\leq \sup_{\substack{w_l \in X_l \\ w_l \neq 0}} \frac{\|[w_l]\|_{0;S} \|\mu_l\|_{0;S}}{\|w_l\|_0} \leq \frac{C}{h_l^2} \|\mu_l\|_{h_l^{-\frac{3}{2}};S} \ .$$

To establish the lower bound, we observe that

$$\|B_l\mu_l\|_0 \geq \frac{b(w_l,\mu_l)}{\|w_l\|_0} \ , \quad w_l \in X_l, \ w_l \neq 0 \ .$$

Each $w_l \in X_l$ is uniquely defined by its values at the vertices of the triangulation. We define $w_l(p) := \mu_l(p)$ for an interior vertex p of a non-mortar side γ_m, $1 \leq m \leq M$. For all other vertices q of $\mathcal{T}_l$, we set $w_l(q) := 0$. This special choice yields $b(w_l,\mu_l) \geq c\|[w_l]\|_{0;\mathcal{S}}\|\mu_l\|_{0;\mathcal{S}}$; we refer to [Woh99c] for details. The lower bound in (2.69) now follows from $\|[w_l]\|_{0;\mathcal{S}} \geq ch_l^{-1/2}\|w_l\|_0$. □

Remark 2.39. *Using a similar construction as in the proof of Lemma 2.38, we find by a straightforward computation and by means of (2.69)*

$$\inf_{\substack{v_l \in X_l \\ B_l^* v_l = B_l^* w_l}} \|v_l\|_0 \leq Ch_l^2\|B_l^* w_l\|_{h_l^{-\frac{3}{2}};\mathcal{S}}, \quad w_l \in X_l \ . \tag{2.70}$$

Using the definition (2.67) of the operator K_l, we can rewrite the saddle point problem (1.9) on $X_l \times M_l$ as an operator equation: Find $z_l^* := (u_l, \lambda_l) \in X_l \times M_l$ such that

$$K_l z_l^* = f_l \ ,$$

where $f_l := (\widetilde{f}_l, 0) \in X_l \times M_l$ is defined by $(\widetilde{f}_l, v_l)_0 := (f, v_l)_0$, $v_l \in X_l$.

To establish appropriate smoothing properties, we will use suitable operator norms. We recall that if S is a linear continuous operator $S : H_1 \longrightarrow H_2$, where H_1 and H_2 are Hilbert spaces with norms $\|\cdot\|_{H_1}$ and $\|\cdot\|_{H_2}$, respectively. Then the standard operator norm is given by

$$\|S\| := \sup_{\substack{x \in H_1 \\ x \neq 0}} \frac{\|Sx\|_{H_2}}{\|x\|_{H_1}} \ .$$

In the next two subsubsections, we establish smoothing and stability properties for two different types of iterations. In particular, we do not require that the iterates satisfy the weak continuity conditions at the interfaces exactly. As a consequence, we neither need a good preconditioner for the exact Schur complement nor an exact solver for a modified Schur complement.

2.5.2.1 A Block Diagonal Smoother. Following the ideas of [Ver84], we introduce a smoother for the squared positive definite system. The operator $\widehat{K}_l : X_l \times M_l \longrightarrow X_l \times M_l$ is defined by means of the symmetric positive definite bilinear forms $\widehat{a}(\cdot,\cdot)$ and $d(\cdot,\cdot)$ on $X_l \times X_l$ and $M_l \times M_l$, respectively,

$$(\widehat{K}_l(v_l, \mu_l), (w_l, \nu_l))_{h_l;\Omega \times \mathcal{S}} := \widehat{a}(v_l, w_l) + d(\nu_l, \mu_l), \quad (w_l, \nu_l) \in X_l \times M_l \ . \tag{2.71}$$

It has a block diagonal structure

$$\widehat{K}_l(v_l, \mu_l) = (\widehat{A}_l v_l, D_l \mu_l) \ ,$$

where the operators $\widehat{A}_l : X_l \longrightarrow X_l$, and $D_l : M_l \longrightarrow M_l$ are associated with the bilinear forms $\widehat{a}(\cdot, \cdot)$ and $d(\cdot, \cdot)$, respectively. One smoothing iteration on level l is given by

$$z_l^m := z_l^{m-1} + \widehat{K}_l^{-1} K_l \widehat{K}_l^{-1}(d_l - K_l z_l^{m-1}), \quad m \geq 1 , \qquad (2.72)$$

where d_l represents the right hand side of the system $K_l z_l = d_l$, which has to be solved, z_l is the exact solution, z_l^m denotes the iterate in the mth-step, and z_l^0 is the initial guess. The block diagonal smoother works on the squared system which is positive definite. Each smoothing step can be easily performed provided that the applications of $\widehat{A}_l^{-1}$ and D_l^{-1} are cheap. The following lemma gives the smoothing rate and can be found in [Woh00b].

Lemma 2.40. (Smoothing property)
Let $\widehat{K}_l$ be defined as in (2.71), where $\widehat{A}_l, \|\widehat{A}_l\| \leq C/h_l^2$, and $D_l, \|D_l\| \leq C/h_l^2$, are self-adjoint positive definite operators. If there exists for each $w_l \in X_l$ an $\alpha_{w_l}, 0 < \alpha_{w_l} < 1$ such that

$$(A_l w_l, w_l)_0 \leq \alpha_{w_l}(\widehat{A}_l w_l, w_l)_0, \quad (B_l D_l^{-1} B_l^* w_l, w_l)_0 \leq (1 - \alpha_{w_l})(\widehat{A}_l w_l, w_l)_0 ,$$

then the following smoothing property holds for the iteration (2.72):

$$\|K_l e_l^m\|_{h_l; \Omega \times S} \leq \frac{C}{h_l^2 \sqrt{m}} \|e_l^0\|_{h_l; \Omega \times S}, \quad m \geq 1 . \qquad (2.73)$$

Here, $e_l^m := z_l^m - z_l, m \geq 0$, is the iteration error in the mth-smoothing step, and the constant C does not depend on the α_{w_l}.

Proof. The iteration error e_l^m satisfies

$$e_l^m = (\mathrm{Id} - \widehat{K}_l^{-1} K_l \widehat{K}_l^{-1} K_l)^m e_l^0, \quad m \geq 1 .$$

Since $\widehat{K}_l$ is a self-adjoint positive definite operator and K_l is self-adjoint, there exists a complete set of orthogonal eigenfunctions z_l^i satisfying

$$\widehat{K}_l^{-\frac{1}{2}} K_l \widehat{K}_l^{-\frac{1}{2}} z_l^i = s_i z_l^i .$$

Setting $(w_l^i, \mu_l^i) := \widehat{K}_l^{-\frac{1}{2}} z_l^i$, we find for the eigenvalues $s_i \neq 0$, that $w_l^i \neq 0$, and that

$$A_l w_l^i + \frac{1}{s_i} B_l D_l^{-1} B_l^* w_l^i = s_i \widehat{A}_l w_l^i .$$

Then, the assumptions on $\widehat{A}_l$ and D_l yield

$$|s_i| \leq \frac{(B_l D_l^{-1} B_l^* w_l^i, w_l^i)_0 + \sqrt{(B_l D_l^{-1} B_l^* w_l^i, w_l^i)_0^2 + 4(\widehat{A}_l w_l^i, w_l^i)_0 (A_l w_l^i, w_l^i)_0}}{2(\widehat{A}_l w_l^i, w_l^i)_0}$$

$$\leq \tfrac{1}{2}((1 - \alpha_{w^i}) + (1 + \alpha_{w^i})) = 1 ,$$

and the norm of $K_l e_l^m$ is bounded by

$$\|K_l e_l^m\|_{h_l;\Omega \times S} \leq \|\widehat{K}_l^{\frac{1}{2}} \widehat{K}_l^{-\frac{1}{2}} K_l \widehat{K}_l^{-\frac{1}{2}} (\text{Id} - (\widehat{K}_l^{-\frac{1}{2}} K_l \widehat{K}_l^{-\frac{1}{2}})^2)^m \widehat{K}_l^{\frac{1}{2}} e_l^0\|_{h_l;\Omega \times S}$$

$$\leq \sup_{s \in \sigma(\widehat{K}_l^{-\frac{1}{2}} K_l \widehat{K}_l^{-\frac{1}{2}})} |s(1-s^2)^m| \, \|\widehat{K}_l\| \, \|e_l^0\|_{h_l;\Omega \times S} \ .$$

We obtain (2.73) by using that $\sup_{t \in [0;1]}(t(1-t^2)^m) \leq C/\sqrt{m}$ and $\|\widehat{K}_l\| \leq C/h_l^2$. $\square$

Combining the approximation property (2.68) and the smoothing property (2.73), we obtain level independent convergence rates for the two-grid algorithm provided that the number of smoothing steps is large enough. The analysis of the full multigrid cycle is based on the two-grid case, a perturbation argument and the stability of the smoothing iteration (2.72); see [Hac85].

Lemma 2.41. (Stability estimate)
Under the assumptions of Lemma 2.40 and if furthermore $\|\widehat{A}_l^{-1}\| \leq Ch_l^2$, then there exists a constant C independent of m such that the following stability estimate holds

$$\|e_l^m\|_{h_l;\Omega \times S} \leq C\|e_l^0\|_{h_l;\Omega \times S}, \quad m \geq 1 \ . \tag{2.74}$$

Proof. To obtain the stability estimate (2.74), we use the same type of arguments as in the proof of Lemma 2.40. The assumption on D_l and (2.70) yield an upper bound for $\|D_l^{-1/2}\|$

$$\|D_l^{-\frac{1}{2}}\| = \sup_{\substack{\mu_l \in M_l \\ \mu_l \neq 0}} \frac{\|D_l^{-\frac{1}{2}}\mu_l\|_{h_l^{-3/2};S}}{\|\mu_l\|_{h_l^{-3/2};S}} = \sup_{\substack{\mu_l \in M_l \\ B_l^* w_l = \mu_l \neq 0}} \frac{\|D_l^{-\frac{1}{2}} B_l^* w_l\|_{h_l^{-3/2};S}}{\|B_l^* w_l\|_{h_l^{-3/2};S}}$$

$$\leq \sup_{\substack{\mu_l \in M_l \\ \mu_l \neq 0}} \inf_{\substack{w_l \in X_l \\ B_l^* w_l = \mu_l}} \frac{\|\widehat{A}_l^{\frac{1}{2}} w_l\|_0}{\|B_l^* w_l\|_{h_l^{-3/2};S}} \leq Ch_l^{-1} h_l^2 \leq Ch_l \ .$$

The last inequality, together with the assumption on $\widehat{A}_l^{-1}$, gives $\|\widehat{K}_l^{-1/2}\| \leq h_l$, and thus

$$\|e_l^m\|_{h_l;\Omega \times S} \leq \|(\text{Id} - \widehat{K}_l^{-1} K_l \widehat{K}_l^{-1} K_l)^m\| \, \|e_l^0\|_{h_l;\Omega \times S}$$

$$\leq \sup_{s \in [-1;1]} (1-s^2)^m \|\widehat{K}_l^{\frac{1}{2}}\| \, \|\widehat{K}_l^{-\frac{1}{2}}\| \, \|e_l^0\|_{h_l;\Omega \times S} \leq C\|e_l^0\|_{h_l;\Omega \times S} \ .$$

$\square$

Under the assumptions of Lemma 2.41, the convergence rates of the $\mathcal{W}$-cycle in the $\|\cdot\|_{h_l;\Omega \times S}$-norm are independent of the number of refinement levels provided that the number of smoothing steps is large enough; see, e.g., [Hac85, Ver84].

Remark 2.42. *A suitable smoother, in the algebraic formulation of the method, is given by the diagonal matrix*

$$\widehat{K}_l := \alpha \, h^{d-2} \begin{pmatrix} \mathrm{Id} & 0 \\ 0 & h_l^2 \mathrm{Id} \end{pmatrix} \, ,$$

for some constant $\alpha > 0$. Here, nodal basis functions are used for both the finite elements and the Lagrange multiplier, and d stands for the space dimension, $\Omega \subset \mathbb{R}^d$.

One iteration step using (2.72) requires the application of $\widehat{K}_l^{-1}$ twice. This is closely related to the fact that $\widehat{K}_l$ is positive definite whereas K_l is indefinite. In the following subsubsection, we discuss a second type of smoothing operator originally analyzed in [Zul01] for an abstract saddle point problem. It has been applied to the mortar setting in [WW99].

2.5.2.2 An Indefinite Smoother. The symmetric but indefinite operator K_l can be decomposed as follows

$$K_l = \begin{pmatrix} A_l & 0 \\ B_l^* & \mathrm{Id} \end{pmatrix} \begin{pmatrix} A_l^{-1} & 0 \\ 0 & -B_l^* A_l^{-1} B_l \end{pmatrix} \begin{pmatrix} A_l & B_l \\ 0 & \mathrm{Id} \end{pmatrix} \, ,$$

where A_l^{-1} has to be replaced by a suitable pseudo-inverse if A_l is singular.

This decomposition motivates the construction of our second smoothing operator $\widetilde{K}_l$. We note that a smoother with the same algebraic structure as K_l was introduced and analyzed in [BS97] for the Stokes problem. In the definition of the smoother, the operator A_l is replaced by a suitable $\widetilde{A}_l$. To apply one smoothing step, one has to solve a modified Schur complement system exactly, where the Schur complement is defined by $S_l := B_l^* \widetilde{A}_l^{-1} B_l$. This approach has been applied successfully to the mortar situation in [BD98, BDW99, Bra01, WW98]. A disadvantage of this approach is that the exact solution of the modified Schur complement system can be rather expensive. Even the use of dual Lagrange multipliers does not, in general, improve the complexity. A simplified approach has been proposed in [Zul01]

$$\widetilde{K}_l := \begin{pmatrix} \widetilde{A}_l & B_l \\ B_l^* & S_l - \widetilde{S}_l \end{pmatrix} = \begin{pmatrix} \widetilde{A}_l & 0 \\ B_l^* & \mathrm{Id} \end{pmatrix} \begin{pmatrix} \widetilde{A}_l^{-1} & 0 \\ 0 & -\widetilde{S}_l \end{pmatrix} \begin{pmatrix} \widetilde{A}_l & B_l \\ 0 & \mathrm{Id} \end{pmatrix} \, . \tag{2.75}$$

Then, the smoothing iteration is defined in terms of $\widetilde{K}_l$

$$z_l^m := z_l^{m-1} + \widetilde{K}_l^{-1}(d_l - K_l z_l^{m-1}) \, , \tag{2.76}$$

where d_l stands for the right hand side of the system $K_l z_l = d_l$ which has to be solved, z_l is the exact solution, z_l^m the iterate in the mth-step, and z_l^0 the initial guess. Each smoothing step can be performed easily provided that the applications of $\widetilde{A}_l^{-1}$ and $\widetilde{S}_l^{-1}$ are cheap. The following lemma has been established in [Zul01], and guarantees the smoothing property of the iteration (2.76) under some assumptions on $\widetilde{S}_l$ and $\widetilde{A}_l$.

Lemma 2.43. (Smoothing property)

Let $\widetilde{K}_l$ be defined as in (2.75), with $\widetilde{A}_l$ and $\widetilde{S}_l$ positive definite self-adjoint operators. Under the assumptions $(A_l w_l, w_l)_0 \leq (\widetilde{A}_l w_l, w_l)_0$, $w_l \in X_l$, and

$$(\widetilde{S}_l \mu_l, \mu_l)_{h_l^{-\frac{3}{2}};S} \leq (S_l \mu_l, \mu_l)_{h_l^{-\frac{3}{2}};S} < \frac{4}{3}(\widetilde{S}_l \mu_l, \mu_l)_{h_l^{-\frac{3}{2}};S}, \quad \mu_l \in M_l , \quad (2.77)$$

we obtain the following smoothing property for the iteration (2.76)

$$\|K_l e_l^m\|_{h_l;\Omega \times S} \leq C \, \eta(m) \, \|\widetilde{K}_l - K_l\| \, \|e_l^0\|_{h_l;\Omega \times S} .$$

Here $\eta(m) \longrightarrow 0$ for $m \longrightarrow \infty$.

We refer to [Zul01] for a proof, and note that the assumption (2.77) can be weakened if a damping strategy is used. A central point in the proof is that the operator $\widetilde{K}_l - K_l$ is positive semi-definite.

In the previous subsubsection, we found it easy to construct a scaled Jacobi-type operator $\widehat{K}_l$ satisfying the assumptions of Lemmas 2.40 and 2.41. Here, the choice of $\widetilde{S}_l$ is a delicate matter. To find an adequate $\widetilde{S}_l$ satisfying (2.77), we follow an approach proposed in [WW99].

The operator $\widetilde{S}_l$ is constructed in terms of a positive definite self-adjoint operator $\widehat{S}_l$ satisfying $S_l < 2\widehat{S}_l$. Then, the spectral radius of $\mathrm{Id} - \widehat{S}_l^{-1} S_l$ is bounded by one, i.e., $q := \rho(\mathrm{Id} - \widehat{S}_l^{-1} S_l) < 1$. With a function $k(\varepsilon)$ defined by

$$k(\varepsilon) := \frac{\log \varepsilon}{\log q}, \quad \varepsilon > 0 ,$$

we find that $\rho((\mathrm{Id} - \widehat{S}^{-1} S)^k) < \varepsilon$ for any integer $k > k(\varepsilon)$. Based on these preliminary remarks, we define

$$\widetilde{S}_l(k, \alpha) := \alpha S_l^{1/2} \left(\mathrm{Id} - (\mathrm{Id} - S_l^{1/2} \widehat{S}_l^{-1} S_l^{1/2})^k \right)^{-1} S_l^{1/2} .$$

A straightforward computation shows that $\widetilde{S}_l(k, \alpha)$ is selfadjoint and positive definite for any integer $k > 0$ and $\alpha > 0$. The idea is now to find an integer k and a value of α such that $\widetilde{S}_l(k, \alpha)$ satisfies (2.77). It is easy to see that

$$\frac{1 - \varepsilon}{\alpha} \widetilde{S}_l(k, \alpha) \leq S_l \leq \frac{1 + \varepsilon}{\alpha} \widetilde{S}_l(k, \alpha)$$

for $k > k(\varepsilon)$. Setting $\alpha := (1 - \varepsilon)$ yields

$$\widetilde{S}_l(k, 1 - \varepsilon) \leq S_l \leq \frac{1 + \varepsilon}{1 - \varepsilon} \widetilde{S}_l(k, 1 - \varepsilon) = (1 + \frac{2\varepsilon}{1 - \varepsilon}) \widetilde{S}_l(k, 1 - \varepsilon) .$$

The last inequality shows that the assumptions of Lemma 2.43 are satisfied for $7\varepsilon < 1$. We define $\widetilde{S}_l := \widetilde{S}_l(k_\varepsilon, 1 - \varepsilon)$ for a fixed $\varepsilon < 1/7$, $k_\varepsilon > k(\varepsilon)$, and refer to Subsect. 2.5.3 for details on the implementation of the matrix vector multiplication of $\widetilde{S}_l^{-1}$.

To obtain optimal convergence rates, not only for the two-grid algorithm but also for the full multigrid method, a stability estimate for the smoothing iteration (2.76) is required.

Lemma 2.44. (Stability estimate)
Let the assumptions of Lemma 2.43 be satisfied. Furthermore, if $ch_l^2\|w_l\|_0 \leq \|\widetilde{A}_l^{-1}w_l\|_0 \leq Ch_l^2\|w_l\|_0$, there exists a constant independent of m such that the following estimate holds

$$\|e_l^m\|_{h_l;\Omega\times S} \leq C\,\|e_l^0\|_{h_l;\Omega\times S}, \quad m \geq 1\ .$$

Proof. The assumptions on $\widetilde{A}_l$ and $\widetilde{S}_l$ yield that $\widetilde{K}_l - K_l$ is a positive semi-definite operator. Thus, we find for e_l^m

$$e_l^m = (\mathrm{Id} - \widetilde{K}_l^{-1}K_l)^m e_l^0 = \widetilde{K}_l^{-1}(\widetilde{K}_l - K_l)\left(\widetilde{K}_l^{-1}(\widetilde{K}_l - K_l)\right)^{m-1} e_l^0$$
$$= \widetilde{K}_l^{-1}(\widetilde{K}_l - K_l)^{1/2}\left((\widetilde{K}_l - K_l)^{1/2}\widetilde{K}_l^{-1}(\widetilde{K}_l - K_l)^{1/2}\right)^{m-1}(\widetilde{K}_l - K_l)^{1/2}e_l^0.$$

Under the assumptions of Lemma 2.43, the spectral radius of $(\widetilde{K}_l - K_l)^{1/2}\widetilde{K}_l^{-1}(\widetilde{K}_l - K_l)^{1/2}$ is bounded by one; see [Zul01], and we obtain

$$\|e_l^m\|_{h_l;\Omega\times S} \leq \|\widetilde{K}_l^{-1}\|\,\|\widetilde{K}_l - K_l\|\,\|e_l^0\|_{h_l;\Omega\times S}$$
$$\leq C\,h_l^{-2}\,h_l^2\|e_l^0\|_{h_l;\Omega\times S} \leq C\,\|e_l^0\|_{h_l;\Omega\times S}\ .$$

In the second inequality, we have used that under the assumptions of Lemma 2.44, the norm of $\widetilde{K}_l^{-1}$ is bounded from above by Ch_l^2 and that of $\widetilde{K}_l - K_l$ by Ch_l^{-2}.   □

Remark 2.45. *In Lemma 2.44, it was assumed that the condition number of $\widetilde{A}_l$ is bounded independently of the meshsize h_l. This is satisfied for Jacobi-type smoothers, but not for those of ILU-type. Nevertheless, level independent convergence rates can be obtained by replacing the mesh dependent norm for the Lagrange multiplier by a norm involving the Schur complement.*

We can now formulate the central result of this section which shows that both classes of smoothers give rise to optimal multigrid methods.

Theorem 2.46. *Let the smoothing iteration (2.6) in the multigrid cycle be defined by (2.72) or (2.76). Then under the assumptions of Lemma 2.41 or Lemma 2.44, the convergence rates of the W-cycle are independent of the number of refinement levels, provided that the number of smoothing steps is large enough.*

Proof. The abstract multigrid theory; see, e.g., [BS94, Hac85, Ver84], shows that the approximation property given in Lemma 2.37, the smoothing properties given in Lemmas 2.40 and 2.43, and the stability estimates of Lemmas 2.41 and 2.44 guarantee level independent convergence rates of the W-cycle provided that the number of smoothing steps is large enough.   □

2.5.3 Numerical Results

We show only some numerical results which were originally presented in [WW99]. In contrast to [BD98, BDW99, Bra01, WW98], where an exact modified Schur complement was solved in each smoothing step, we do not solve any modified Schur complement systems exactly. As previously shown, the exact solution can be replaced by an iteration. As a consequence the iterates do not belong to the subspace for which the saddle point problem is positive definite. In spite of this, the resulting multigrid convergence rates are independent of the refinement levels provided that the number of smoothing steps is large enough.

We start with a discussion of a good choice of $\widetilde{S}_l$ in (2.75). We have shown that $\widetilde{S}_l := \widetilde{S}_l(k_\varepsilon, 1 - \varepsilon)$ satisfies the assumptions of Lemma 2.43 provided that $\varepsilon < 1/7$, $S_l < 2\widehat{S}_l$, and $k_\varepsilon > k(\varepsilon)$. We select $\varepsilon := 0.1$ in our numerical examples, and we obtain the solution of $\widetilde{S}_l y_l = t_l$ by k_ε iteration steps and one scaling step. Formally, we can rewrite $\widetilde{S}_l y_l = t_l$ as

$$y_l = \frac{1}{1 - \varepsilon} S_l^{-1/2} \left(\mathrm{Id} - (\mathrm{Id} - S_l^{1/2} \widehat{S}_l^{-1} S_l^{1/2})^{k_\varepsilon} \right) S_l^{-1/2} t_l .$$

The implementation of $\widetilde{S}_l^{-1} t_l$ is based on the identity

$$(1 - \varepsilon) \widetilde{S}_l^{-1} S_l = \mathrm{Id} - (\mathrm{Id} - \widehat{S}_l^{-1} S_l)^{k_\varepsilon} ,$$

which is established by a straightforward computation. The right hand side can be interpreted as the error propagation of the following iteration scheme:

$$y_l^n := y_l^{n-1} + \widehat{S}_l^{-1}(t_l - S_l y_l^{n-1}), \quad n \geq 1 . \tag{2.78}$$

Setting $y_l^0 := 0$, we find $y_l^n - S_l^{-1} t_l = -(\mathrm{Id} - \widehat{S}_l^{-1} S_l)^n S_l^{-1} t_l$. The choice $n = k_\varepsilon$ yields $y_l = (1 - \varepsilon)^{-1} y_l^{k_\varepsilon}$. The delicate point is the choice of k_ε. In our approach, we use a stopping criteria for the iteration (2.78) to define k_ε.

The application of $\widetilde{K}_l^{-1}$ in each smoothing step can be easily implemented in terms of the inner iteration (2.78). Each smoothing step (2.6) requires the solution of

$$\widetilde{K}_l \begin{pmatrix} x_l \\ y_l \end{pmatrix} = \begin{pmatrix} f_l \\ g_l \end{pmatrix} .$$

Using the decomposition (2.75), we find that the inverse of $\widetilde{K}_l$ can be written as a product of an upper and a lower block tridiagonal matrix

$$\widetilde{K}_l^{-1} = \begin{pmatrix} \mathrm{Id} & -\widetilde{A}_l^{-1} B_l \widetilde{S}_l^{-1} \\ & \widetilde{S}_l^{-1} \end{pmatrix} \begin{pmatrix} \widetilde{A}_l^{-1} & 0 \\ B_l^* \widetilde{A}_l^{-1} & -\mathrm{Id} \end{pmatrix} =: U_l L_l .$$

Then, the solution can be obtained by a forward and backward substitution, i.e., $(x_l^T, y_l^T)^T = U_l(s_l^T, t_l^T)^T$, where $(s_l^T, t_l^T)^T = L_l(f_l^T, g_l^T)^T$. The application of $\widetilde{K}_l^{-1}$ is carried out in the following way:

$$s_l := \widetilde{A}_l^{-1} f_l \ ,$$

$$t_l := B_l^* s_l - g_l \ , \qquad y_l^0 := 0 \ ,$$

$$\text{for } n = 1, 2, 3, \ldots \text{ do}$$

$$y_l^n := y_l^{n-1} + \widehat{S}_l^{-1}(t_l - B_l^* \widetilde{A}_l^{-1} B_l y_l^{n-1}) \ , \qquad (2.79)$$

$$\text{until } \|t_l - B_l^* \widetilde{A}_l^{-1} B_l y_l^n\|_{h_l^{-3/2};S} \le \varepsilon \|t_l\|_{h_l^{-3/2};S} \ ,$$

$$y_l := \frac{1}{1 - \varepsilon} \, y_l^n \ ,$$

$$x_l := s_l - \widetilde{A}_l^{-1} B_l y_l \ .$$

The linear iteration (2.79) can be accelerated by a conjugate gradient method. In our numerical results only a few number of inner iteration steps were required.

In our first example, we consider a problem with highly discontinuous coefficients similar to Example 3 in Subsect. 1.5.1. The domain is decomposed into four squares and the coefficient a is 1 or 10^6 in the subdomains. For details, we refer to [WW99] where all the examples of this subsection were originally discussed. This problem is a classical test example in multilevel theory; see, e.g., [Den82]. Our numerical results are based on the smoother given by (2.79).

Table 2.9. Asymptotic convergence rates for highly discontinuous coefficients

level	number of elements	$\mathcal{W}(2,2)$-cycle damped Jacobi	$\mathcal{V}(1,1)$-cycle symmetric Gauß–Seidel
4	1024	0.084	0.080
5	4096	0.128	0.091
6	16384	0.137	0.095
7	65536	0.144	0.098
8	262144	0.146	0.102

Two choices of $\widetilde{A}_l$ are considered. The first one is a Jacobi method with damping factor 0.7 and the second one is a symmetric Gauß–Seidel smoother. In Table 2.9, we present the asymptotic convergence rates for a $\mathcal{W}$-cycle with two pre- and postsmoothing steps and a $\mathcal{V}$-cycle with one pre- and postsmoothing step. We find that the number of inner iteration steps required in (2.79) is bounded independently of the refinement level, and by 4 for the Jacobi-type smoother and by 8 for the Gauß–Seidel smoother. This reflects the fact that the condition number of the approximated Schur complement is worse for the Gauß–Seidel smoother. Within one smoothing step (2.79), we have to apply $\widetilde{A}_l^{-1}$ $(n+2)$-times, where n is the number of inner iterations. At first glance, this makes the Gauß–Seidel smoother considerably more expensive than the Jacobi smoother. However, the application of $\widetilde{A}_l^{-1} B_l$ can be extremely simplified by taking the structure of B_l into account.

In the rest of this subsection, we have also applied the multigrid method for the examples given in the introduction of Sect. 1.5. We use the $\mathcal{V}$-cycle as a preconditioner for a Krylov space method. Since we are not working on the subspace on which the operator K_l is positive definite, we use a bicgstab method. An ILU-type smoother is chosen in the example of the time-dependent problem illustrated in Fig. 1.25. We recall that in the case of a nested iteration, the reduction factor is much better at the beginning. Only a few iterations steps are necessary to obtain an iteration error of the same order as the discretization error. An error reduction of 10^{-10} is obtained with 3 iteration steps on each level. On the finest mesh, we have 327680 elements.

The multigrid method is more sensitive to the choice of the smoother in the case when mixed and conforming finite elements are coupled, as analyzed in Sect. 1.4.1 and implemented for the example illustrated by Fig. 1.26 in Sect. 1.5. Numerical tests with a Jacobi-type smoother show that the damping factor has to be decreased and that the number of smoothing steps has to be increased to obtain a robust method. Stable convergence rates are obtained by using a preconditioned Krylov space method. We use a $\mathcal{V}(2, 2)$-cycle with an ILU smoother and three inner iterations in (2.79) as preconditioner. Table 2.10 shows the performance of the preconditioned method.

Table 2.10. Convergence rate of the preconditioned Krylov space method

level	elements	conv. rate
1	4028	0.05
2	16112	0.08
3	64448	0.10
4	257792	0.11

In the case of the linear elasticity problem discussed in Sect. 1.5; see also Fig. 1.27, the multigrid method is very sensitive to large aspect ratios of the subdomains and to the material parameters. The average convergence rate of the preconditioned bicgstab is 0.5 in our numerical experiments. A $\mathcal{V}(3, 3)$-cycle with a symmetric Gauß–Seidel smoother is used as preconditioner. On the finest mesh, we have 360448 elements.

Bibliography

[AB85] D.N. Arnold and F. Brezzi. Mixed and nonconforming finite element methods: Implementation, post-processing and error estimates. M^2AN Math. Modelling Numer. Anal., 19:7–35, 1985.

[AFW97] D.N. Arnold, R.S. Falk, and R. Winther. Preconditioning in $H(\text{div})$ and applications. Math. Comput., 66:957–984, 1997.

[AFW98] D.N. Arnold, R.S. Falk, and R. Winther. Multigrid preconditioning in $H(\text{div})$ on non-convex polygons. Comput. and Appl. Math., 17:303–315, 1998.

[AFW00] D.N. Arnold, R.S. Falk, and R. Winther. Multigrid in H(div) and H(curl). Numer. Math., 85:197–217, 2000.

[AK95] Y. Achdou and Y. Kuznetsov. Substructuring preconditioners for finite element methods on nonmatching grids. East-West J. Numer. Math., 3:1–28, 1995.

[AKP95] Y. Achdou, Y. Kuznetsov, and O. Pironneau. Substructuring preconditioners for the Q_1 mortar element method. Numer. Math., 71:419–449, 1995.

[AMW96] Y. Achdou, Y. Maday, and O.B. Widlund. Méthode iterative de sous-structuration pour les éléments avec joints. C. R. Acad. Sci., Paris, Ser. I, 322:185–190, 1996.

[AMW99] Y. Achdou, Y. Maday, and O.B. Widlund. Iterative substructuring preconditioners for mortar element methods in two dimensions. SIAM J. Numer. Anal., 36:551–580, 1999.

[AT95] A. Agouzal and J.M. Thomas. Une methode d'elements finis hybrides en decomposition de domaines. RAIRO Mathematical Modelling and Numerical Analysis, 29:749–764, 1995.

[AV99] A. Alonso and A. Valli. An optimal domain decomposition preconditioner for low-frequency time-harmonic Maxwell equations. Math. Comp., 68:607–631, 1999.

[AY97] T. Arbogast and I. Yotov. A non-mortar mixed finite element method for elliptic problems on non-matching multiblock grids. Comput. Meth. Appl. Mech. Eng., 149:255–265, 1997.

[Bae91] E. Baensch. Local mesh refinement in 2 and 3 dimensions. IMPACT Comput. Sci. Eng., 3:181–191, 1991.

[Ban96] R.E. Bank. Hierarchical bases and the finite element method. Acta Numerica, 5:1–43, 1996.

[BBJ+97] P. Bastian, K. Birken, K. Johannsen, S. Lang, N. Neuß, H. Rentz–Reichert, and C. Wieners. UG – a flexible software toolbox for solving partial differential equations. Computing and Visualization in Science, 1:27–40, 1997.

[BD98] D. Braess and W. Dahmen. Stability estimates of the mortar finite element method for 3-dimensional problems. East-West J. Numer. Math., 6:249–263, 1998.

[BDH⁺99a] R. Beck, P. Deuflhard, R. Hiptmair, R.H.W. Hoppe, and B. Wohlmuth. Adaptive multilevel methods for edge element discretizations of Maxwell's equations. *Surv. Math. Ind.*, 8:271–312, 1999.

[BDH99b] D. Braess, M. Dryja, and W. Hackbusch. Multigrid method for nonconforming fe-discretisations with application to nonmatching grids. *Computing*, 63:1–25, 1999.

[BDL99] D. Braess, P. Deuflhard, and K. Lipnikov. A cascadic conjugate gradient method for domain decomposition with non-matching grids. *Preprint SC99–07, Konrad–Zuse–Zentrum für Informationstechnik Berlin*, 1999.

[BDW99] D. Braess, W. Dahmen, and C. Wieners. A multigrid algorithm for the mortar finite element method. *SIAM J. Numer. Anal.*, 37:48–69, 1999.

[Ben99] F. Ben Belgacem. The mortar finite element method with Lagrange multipliers. *Numer. Math.*, 84:173–197, 1999.

[Bey95] J. Bey. Tetrahedral grid refinement. *Computing*, 55:355–378, 1995.

[BF91] F. Brezzi and M. Fortin. *Mixed and hybrid finite element methods*. Springer–Verlag, New York, 1991.

[BFMR98] F. Brezzi, L. Franca, D. Marini, and A. Russo. Stabilization techniques for domain decomposition methods with non-matching grids. In P. Bjørstad, M. Espedal, and D. Keyes, editors, *Proceedings of the 9th International Conference on Domain Decomposition*, pages 1–11, Bergen, 1998. Domain Decomposition Press.

[BGLV89] J.-F. Bourgat, R. Glowinski, P. Le Tallec, and M. Vidrascu. Variational formulation and algorithm for trace operator in domain decomposition calculations. In T. Chan, R. Glowinski, J. Périaux, and O. Widlund, editors, *Domain Decompositions Methods*, pages 3–16. SIAM, Philadelphia, 1989.

[BH83] D. Braess and W. Hackbusch. A new convergence proof for the multigrid method including the $\mathcal{V}$-cycle. *SIAM J. Numer. Anal.*, 20:967–975, 1983.

[BH99] R. Becker and P. Hansbo. A finite element method for domain decompositions with non-matching grids. *Preprint 3613, INRIA, Sophia Antipolis*, 1999.

[BM95] C. Bernardi and Y. Maday. Raffinement de maillage en elements finis par la methode des joints. *C. R. Acad. Sci., Paris, Ser. I 320*, pages 373–377, 1995. This paper appeared also as a preprint, Laboratoire d'Analyse Numérique, Univ. Pierre et Marie Curie, Paris, R94029, including more details.

[BM97] F. Ben Belgacem and Y. Maday. The mortar element method for three dimensional finite elements. M^2AN, 31:289–302, 1997.

[BM00] F. Brezzi and D. Marini. Error estimates for the three-field formulation with bubble stabilization. *Math. Comp.*, posted on March 24,2000. PII: S0025-5718(00)01250-3 (to appear in print).

[BMP93] C. Bernardi, Y. Maday, and A.T. Patera. Domain decomposition by the mortar element method. In H. Kaper et al., editor, *In: Asymptotic and numerical methods for partial differential equations with critical parameters*, pages 269–286. Reidel, Dordrecht, 1993.

[BMP94] C. Bernardi, Y. Maday, and A.T. Patera. A new nonconforming approach to domain decomposition: the mortar element method. In H. Brezzi et al., editor, *In: Nonlinear partial differential equations and their applications*, pages 13–51. Paris, 1994.

[BPS86a] J.H. Bramble, J.E. Pasciak, and A.H. Schatz. The construction of preconditioners for elliptic problems by substructuring I. *Math. Comp.*, 47:103–134, 1986.

[BPS86b] J.H. Bramble, J.E. Pasciak, and A.H. Schatz. An iterative method for elliptic problems on regions partitioned into substructures. *Math. Comp.*, 46:361–369, 1986.

[BPS89] J.H. Bramble, J.E. Pasciak, and A.H. Schatz. The construction of preconditioners for elliptic problems by substructuring. IV. *Math. Comput.*, 53:1–24, 1989.

[BPWX91a] J.H. Bramble, J.E. Pasciak, J. Wang, and J. Xu. Convergence estimates for multigrid algorithms without regularity assumptions. *Math. Comp.*, 57:23–45, 1991.

[BPWX91b] J.H. Bramble, J.E. Pasciak, J. Wang, and J. Xu. Convergence estimates for product iterative methods with applications to domain decomposition. *Math. Comp.*, 57:1–21, 1991.

[BPX90a] J.H. Bramble, J.E. Pasciak, and J. Xu. Parallel multilevel preconditioners. *Math. Comp.*, 55:1–22, 1990.

[BPX90b] J.H. Bramble, J.E. Pasciak, and J. Xu. Parallel multilevel preconditioners. In T. Chan, R. Glowinski, J. Periaux, and O. Widlund, editors, *Third international symposium on domain decomposition methods for partial differential equations*, pages 341–357, 1990.

[Bra66] J.H. Bramble. A second order finite difference analogue of the first biharmonic boundary value problem. *Numer. Math.*, 9:236–249, 1966.

[Bra93] J.H. Bramble. *Multigrid Methods*. Longman Scientific & Technical, Burnt Mill, Harlow, Essex CM20 2JE, England, 1993. Pitman Research Notes in Mathematics Series #294.

[Bra97] D. Braess. *Finite elements. Theory, fast solvers, and applications in solid mechanics*. Cambridge Univ. Press., 1997.

[Bra01] D. Braess. Analysis of a multigrid algorithm for the mortar finite element method. In *Proceedings of the 12th International Conference on Domain Decomposition*, Chiba, to appear 2001.

[Bre89] S.C. Brenner. An optimal order multigrid method for P1 nonconforming finite elements. *Math. Comp.*, 52:1–15, 1989.

[Bre92] S.C. Brenner. A multigrid algorithm for the lowest order Raviart–Thomas mixed triangular finite element method. *SIAM J. Numer. Anal.*, 29:647–678, 1992.

[Bre00] S.C. Brenner. Lower bounds for two-level additive Schwarz preconditioners with small overlap. *SIAM J. Sci. Comput.*, 21:1657–1669, 2000.

[BS94] S.C. Brenner and L.R. Scott. *The Mathematical Theory of Finite Element Methods*. Springer–Verlag, New York, 1994.

[BS97] D. Braess and R. Sarazin. An efficient smoother for the Stokes problem. *Applied Numer. Math.*, 23:3–19, 1997.

[BS00] S.C. Brenner and L.-Y. Sung. Lower bounds for nonoverlapping domain decomposition preconditioners in two dimensions. *Math. Comp.*, 69:1319–1339, 2000.

[BV90] D. Braess and R. Verfürth. Multigrid methods for nonconforming finite element methods. *SIAM J. Numer. Anal.*, 27:979–986, 1990.

[BW84] P.E. Bjørstad and O.B. Widlund. Solving elliptic problems on regions partitioned into substructures. In G. Birkhoff and A. Schoenstadt, editors, *Elliptic Problem Solvers II*, pages 245–256. Academic Press, New York, 1984.

[BW86] P.E. Bjørstad and O.B. Widlund. Iterative methods for the solution of elliptic problems on regions partitioned into substructures. *SIAM J. Numer. Anal.*, 23:1093–1120, 1986.

[BY93] F. Bornemann and H. Yserentant. A basic norm equivalence for the theory of multilevel methods. *Numer. Math.*, 64:455–476, 1993.

[Cas97] M.A. Casarin. Quasi-optimal Schwarz methods for the conforming spectral element discretization. *SIAM J. Numer. Anal.*, 34:2482–2502, 1997.

[CDS98] X.-C. Cai, M. Dryja, and M. Sarkis. Overlapping nonmatching grid mortar element methods for elliptic problems. *SIAM J. Numer. Anal. 36*, 36:581–606, 1998.

[Cia88] P.G. Ciarlet. *Mathematical Elasticity; Volume 1: Three–Dimensional Elasticity*, volume 20 of *Studies in Mathematics and its Applications*. North-Holland, Amsterdam, 1988.

[CJ97] C. Carstensen and S. Jansche. A posteriori error estimates and adaptive mesh-refining for non-conforming finite element methods. *Berichtsreihe des Mathematischen Seminars Kiel, Preprint 97-8 Universität Kiel*, 1997.

[CJ98] C. Carstensen and S. Jansche. A posteriori error estimates for nonconforming finite element methods. *Z. Angew. Math. Mech.*, 78:S871–S872, 1998.

[CLM97] L. Cazabeau, C. Lacour, and Y. Maday. Numerical quadratures and mortar methods. In *Computational science for the 21st century. Dedicated to Prof. Roland Glowinski on the occasion of his 60th birthday. Symposium, Tours, France, May 5-7, 1997*, pages 119–128. John Wiley & Sons Ltd., 1997.

[CM94] T.F. Chan and T.P. Mathew. Domain decomposition algorithms. *Acta Numerica*, pages 61–143, 1994.

[CPRY97] Z. Cai, R.R. Parashkevov, T.F. Russel, and X. Ye. Domain decomposition for a mixed finite element method in three dimensions. *SIAM J. Numer. Anal.*, 1997. Submitted.

[CR74] M. Crouzeix and P.-A. Raviart. Conforming and nonconforming finite element methods for solving the stationary stokes equations. I. *Revue Franc. Automat. Inform. Rech. operat.*, R-3(7 (1973)):33–76, 1974.

[CW92] X.-C. Cai and O.B. Widlund. Domain decomposition algorithms for indefinite elliptic problems. *SIAM J.Sci. Stat. Comput.*, 13:243–258, 1992.

[CW93] X.-C. Cai and O.B. Widlund. Multiplicative Schwarz algorithms for some nonsymmetric and indefinite problems. *SIAM J. Numer. Anal.*, 30:936–952, 1993.

[CW96] M. Casarin and O.B. Widlund. A hierarchical preconditioner for the mortar finite element method. *ETNA*, 4:75–88, 1996.

[DDPV96] E. Dari, R. Duran, C. Padra, and V. Vampa. A posteriori error estimators for nonconforming finite element methods. M^2AN, 30:385–400, 1996.

[Den82] J.E. Dendy. Black box multi-grid. *J. Comput. Physics*, 48:366–386, 1982.

[DL91] Y.-H. De Roeck and P. Le Tallec. Analysis and test of a local domain decomposition preconditioner. In R. Glowinski, Y. Kuznetsov, G. Meurant, J. Périaux, and O. Widlund, editors, *Fourth International Symposium on Domain Decompositions Methods for Partial Differential Equations*, pages 112–128. SIAM, Philadelphia, 1991.

[Dry88] M. Dryja. A method of domain decomposition for three-dimensional finite element elliptic problems. In R. Glowinski, G. Golub, G. Meurant, and J. Périaux, editors, *Proceedings of the 1st International Conference on Domain Decomposition*, pages 43–61, 1988.

[Dry96] M. Dryja. Additive Schwarz methods for elliptic mortar finite element problems. In K. Malanowski, Z. Nahorski, and M. Peszynska, editors, *Modelling and optimization of distributed parameter systems. Applications to engineering*, pages 31–50. IFIP, Chapman & Hall, London, 1996.

[Dry97] M. Dryja. An iterative substructuring method for elliptic mortar finite element problems with a new coarse space. *East–West J. Numer. Math.*, 5:79–98, 1997.

[Dry98a] M. Dryja. An additive Schwarz method for elliptic mortar finite element problems in three dimensions. In P. Bjørstad, M. Espedal, and D. Keyes, editors, *Proceedings of the 9th International Conference on Domain Decomposition*, pages 88–96, Bergen, 1998. Domain Decomposition Press.

[Dry98b] M. Dryja. An iterative substructuring method for elliptic mortar finite element problems with discontinuous coefficients. In J. Mandel, C. Farhat, and X. Cai, editors, *Proceedings of the 10th International Conference on Domain Decomposition*, pages 94–103. AMS, Contemporary Mathematics series, 1998.

[Dry99] M. Dryja. A Dirichlet–Neumann algorithm for elliptic mortar finite element problems. In W. Hackbusch and S. Sauter, editors, *Numerical Techniques for Composite Materials*, Notes on Numerical Fluid Mechanics. Vieweg, Braunschweig, Submitted to 15th GAMM–Seminar 1999.

[Dry00] M. Dryja. The Dirichlet–Neumann algorithm for mortar saddle point problems. *BIT*, to appear 2000.

[DSW94] M. Dryja, B.F. Smith, and O.B. Widlund. Schwarz analysis of iterative substructuring algorithms for elliptic problems in three dimensions. *SIAM J. Numer. Anal.*, 31:1662–1694, 1994.

[DW94] M. Dryja and O.B. Widlund. Domain decomposition algorithms with small overlap. *SIAM J. Sci. Comput.*, 15:604–620, 1994.

[DW95] M. Dryja and O.B. Widlund. Schwarz methods of Neumann–Neumann type for three-dimensional elliptic finite element problems. *Comm. Pure Appl. Math.*, 48:121–155, 1995.

[EHI+98] B. Engelmann, R.H.W. Hoppe, Y. Iliash, Y. Kuznetsov, Y. Vassilevski, and B.I. Wohlmuth. Adaptive macro-hybrid finite element methods. In H. Bock, F. Brezzi, R. Glowinski, G. Kanschat, Y. Kuznetsov, J. Périaux, and R. Rannacher, editors, *Proc. 2nd European Conference on Numerical Methods*, pages 294–302. World Scientific, Singapore, 1998.

[EHI+00] B. Engelmann, R.H.W. Hoppe, Y. Iliash, Y. Kuznetsov, Y. Vassilevski, and B.I. Wohlmuth. Adaptive finite element methods for domain decompositions on nonmatching grids. In P. Bjørstad and M. Luskin, editors, *Parallel solution of PDEs*, volume 120, pages 57–84. IMA, Springer, Berlin–Heidelberg–New York, 2000.

[EW92] R.E. Ewing and J. Wang. Analysis of the Schwarz algorithm for mixed finite element methods. *RAIRO Mathematical Modelling and Numerical Analysis*, 26:739–756, 1992.

[GC97] B. Guo and W. Cao. Additive Schwarz methods for the h-p version of the finite element method in two dimensions. *SIAM J. Sci. Comput.*, 18:1267–1288, 1997.

[GC98] B. Guo and W. Cao. Additive Schwarz methods for the h-p version of the finite element method in three dimensions. *SIAM J. Numer. Anal.*, 35:632–654, 1998.

[GO95] M. Griebel and P. Oswald. On the abstract theory of additive and multiplicative Schwarz algorithms. *Numer. Math.*, 70:163–180, 1995.

[Gop99] J. Gopalakrishnan. *On the mortar finite element method*. PhD thesis, Texas A&M University, 1999.

[GP00] J. Gopalakrishnan and J.E. Pasciak. Multigrid for the mortar finite element method. *SIAM J. Numer. Anal.*, 37:1029–1052, 2000.

[Gur81] M.E. Gurtin. *An Introduction to Continuum Mechanics*. Academic Press, New York, 1981.

[Hac85] W. Hackbusch. *Multi-Grid Methods and Applications*. Springer, 1985.

[HIK+98] R.H.W. Hoppe, Y. Iliash, Y. Kuznetsov, Y. Vassilevski, and B.I. Wohlmuth. Analysis and parallel implementation of adaptive mortar finite element methods. *East–West J. of Numer. Math.*, 6:223–248, 1998.

[Hip96] R. Hiptmair. *Multilevel Preconditioning for Mixed Problems in Three Dimensions*. PhD thesis, Mathematisches Institut, Universität Augsburg, 1996.

[Hip97] R. Hiptmair. Multigrid method for $H(\mathrm{div})$ in three dimensions. *ETNA*, 6:133–152, 1997.

[Hip98] R. Hiptmair. Multigrid method for Maxwell's equations. *SIAM J. Numer. Anal.*, 36:204–225, 1998.

[HT00] R. Hiptmair and A. Toselli. Overlapping and multilevel Schwarz methods for vector valued elliptic problems in three dimensions. In P. Bjørstad and M. Luskin, editors, *Parallel solution of PDEs*, volume 120, pages 181–208. IMA, Springer, Berlin–Heidelberg–New York, 2000.

[HW97] R.H.W. Hoppe and B.I. Wohlmuth. Adaptive multilevel techniques for mixed finite element discretizations of elliptic boundary value problems. *SIAM J. Numer. Anal.*, 34:1658–1681, 1997.

[KK99] R. Kornhuber and R. Krause. On monotone multigrid methods for the Signorini problem. In W. Hackbusch and S.A. Sauter, editors, *Numerical Techniques for Composite Materials*, Notes on Numerical Fluid Mechanics. Vieweg, Braunschweig, Submitted to 15th GAMM–Seminar 1999.

[KK00] R. Kornhuber and R. Krause. Adaptive multigrid methods for Signorini's problem in linear elasticity. Technical Report A–9, FU Berlin, 2000.

[KLPV00] C. Kim, R.D. Lazarov, J.E. Pasciak, and P.S. Vassilevski. Multiplier spaces for the mortar finite element method in three dimensions. *Preprint, Texas A&M University*, 2000. to appear in SINUM 2001.

[KO88] N. Kikuchi and J.T. Oden. *Contact problems in elasticity: A study of variational inequalities and finite element methods.* SIAM Studies in Applied Mathematics 8, Philadelphia, 1988.

[Kor97] R. Kornhuber. *Adaptive monotone multigrid methods for nonlinear variational problems.* Teubner–Verlag, Stuttgart, 1997.

[Kra01] R.H. Krause. *Monotone Multigrid Methods for Signorini's Problem with Friction.* PhD thesis, FU Berlin, 2001.

[Kuz95a] Y. Kuznetsov. Efficient iterative solvers for elliptic finite element problems on nonmatching grids. *Russ. J. Numer. Anal. Model.*, 10:187–211, 1995.

[Kuz95b] Y. Kuznetsov. Iterative solvers for elliptic finite element problems on nonmatching grids. In *Proc. Int. Conf. AMCA-95*, pages 64–76, Novosibirsk, 1995. NCC publisher.

[Kuz98] Y. Kuznetsov. Overlapping domain decomposition with non-matching grids. In P. Bjørstad, M. Espedal, and D. Keyes, editors, *Proceedings of the 9th International Conference on Domain Decomposition*, pages 606–617, Bergen, 1998. Domain Decomposition Press.

[KW95] Y. Kuznetsov and M.F. Wheeler. Optimal order substructuring preconditioners for mixed finite elements on non-matching grids. *East–West J. Numer. Math.*, 3:127–143, 1995.

[KW00a] R.H. Krause and B.I. Wohlmuth. Domain decomposition methods on nonmatching grids and some applications to linear elasticity problems. *submitted to ZAMM*, 2000.

[KW00b] R.H. Krause and B.I. Wohlmuth. Multigrid methods for mortar finite elements. In E. Dick, K. Riemslagh, and J. Vierendeels, editors, *Multigrid Methods VI*, volume 14 of *Lecture Notes in Computational Science and Engeneering*, pages 136–142, Berlin Heidelberg, 2000. Springer. Proceedings of the Sixth European Multigrid Conference Held in Gent, Belgium, September 27-30, 1999.

[KW00c] R.H. Krause and B.I. Wohlmuth. Nonconforming domain decomposition techniques for linear elasticity. *East–West J. Numer. Math.*, 8:177–206, 2000.

[KW01] R.H. Krause and B.I. Wohlmuth. A Dirchlet–Neumann type algorithm for contact problems with friction. Technical report, Universität Augsburg, 2001.

[Lac98] C. Lacour. Iterative substructuring preconditioners for the mortar finite element method. In P. Bjørstad, M. Espedal, and D. Keyes, editors, *Proceedings of the 9th International Conference on Domain Decomposition*, pages 406–412, Bergen, 1998. Domain Decomposition Press.

[LDV91] P. Le Tallec, Y.-H. De Roeck, and M. Vidrascu. Domain decompositions methods for large linearly elliptic three dimensional problems. *J. Comput. Appl. Math.*, 34, 1991.

[Le 93] P. Le Tallec. Neumann–Neumann domain decomposition algorithms for solving 2D elliptic problems with nonmatching grids. *East-West J. Numer. Math.*, 1:129–146, 1993.

[Le 94] P. Le Tallec. Domain decomposition methods in computational mechanics. In J. Tinsley Oden, editor, *Computational Mechanics Advances*, volume 1 (2), pages 121–220. North–Holland, 1994.

[Lio88] P.-L. Lions. On the Schwarz alternating method I. In R. Glowinski, G. Golub, G. Meurant, and J. Périaux, editors, *First International Symposium on Domain Decomposition Methods for Partial Differential Equations*, pages 1–42, Philadelphia, PA, 1988. SIAM.

[LPV99] R.D. Lazarov, J.E. Pasciak, and P.S. Vassilevski. Iterative solution of a combined mixed and standard Galerkin discretization method for elliptic problems. *Preprint Texas A & M University*, 1999. to appear in Journal of Computational Linear Algebra.

[LSV94] P. Le Tallec, T. Sassi, and M. Vidrascu. Three-dimensional domain decomposition methods with nonmatching grids and unstructured coarse solvers. In D. Keyes et al., editor, *Domain decomposition methods in scientific and engineering computing. Proceedings of the 7th international conference on domain decomposition*, pages 61–74. American Mathematical Society. Contemp. Math. 180, 1994.

[Mat93a] T.P. Mathew. Schwarz alternating and iterative refinement methods for mixed formulations of elliptic problems, Part I: Algorithms and Numerical results. *Numer. Math.*, 65:445–468, 1993.

[Mat93b] T.P. Mathew. Schwarz alternating and iterative refinement methods for mixed formulations of elliptic problems, Part II: Theory. *Numer. Math.*, 65:469–492, 1993.

[McC87] S. McCormick. *Multigrid Methods*. SIAM Frontiers in Applied Mathematics 3, Philadelphia, 1987.

[MH94] J.E. Marsden and T.J.R. Hughes. *Mathematical Foundations of Elasticity*. Dover, 1994. Originally published by Prentice Hall, 1983.

[Néd82] J.C. Nédélec. Elements finis mixtes incompressible pour l'equation de Stokes dans $\mathbb{R}^3$. *Numer. Math.*, 39:97–112, 1982.

[Nit70] J. Nitsche. Über ein Variationsprinzip zur Lösung von Dirichlet Problemen bei Verwendung von Teilräumen, die keinen Randbedingungen unterworfen sind. *Abh. Math. Univ. Hamburg*, 36:9–15, 1970.

[Osw94] P. Oswald. *Multilevel finite element approximation*. Teubner Skripten zur Numerik. B.G. Teubner, Stuttgart, 1994.

[OW00] P. Oswald and B. Wohlmuth. On polynomial reproduction of dual FE bases. Technical Report 10009640-000512-07, Bell Laboratories, Lucent Technologies, 2000.

[Pav94a] L.F. Pavarino. Additive Schwarz methods for the p-version finite element method. *Numer. Math.*, 66:493–515, 1994.

[Pav94b] L.F. Pavarino. Schwarz methods with local refinement for the p-version finite element method. *Numer. Math.*, 69:185–211, 1994.

[PS96] J. Pousin and T. Sassi. Adaptive finite element and domain decomposition with non matching grids. In J. Désidéri et al., editor, *Proc. 2nd ECCOMAS Conf. on Numer. Meth. in Engrg., Paris, September 1996*, pages 476–481. Wiley, Chichester, 1996.

[PW97] L.F. Pavarino and O.B. Widlund. Iterative substructuring methods for spectral elements: Problems in three dimensions based on numerical quadrature. *Computers Math. Applic.*, 33:193–209, 1997.

[PW00a] L.F. Pavarino and O.B. Widlund. Iterative substructuring methods for spectral element discretizations of elliptic systems. I: Compressible linear elasticity. *SIAM J. Numer. Anal.*, 37:353–374, 2000.

[PW00b] L.F. Pavarino and O.B. Widlund. Iterative substructuring methods for spectral element discretizations of elliptic systems. II: Mixed methods for linear elasticity and Stokes flow. *SIAM J. Numer. Anal.*, 37:375–402, 2000.

[QV97] A. Quarteroni and A. Valli. *Numerical approximation of partial differential equations*, volume 23 of *Computational mathematics*. Springer, 1997.

[QV99] A. Quarteroni and A. Valli. *Domain Decomposition Methods for Partial Differential Equations*. Numerical Mathematics and Scientific Computation. Oxford University Press, 1999.

[RT77] P.A. Raviart and J.M. Thomas. A mixed finite element method for 2-nd order elliptic problems. In *Math. Aspects Finite Elem. Meth.*, Lect. Notes Math. 606, pages 292–315, 1977.

[Sar94] M.V. Sarkis. *Schwarz Preconditioners for Elliptic Problems with Discontinuous Coefficients Using Conforming and Non-Conforming Elements*. PhD thesis, Courant Institute, New York University, 1994.

[SBG96] B.F. Smith, P.E. Bjørstad, and W.D. Gropp. *Domain Decomposition: Parallel Multilevel Methods for Elliptic Partial Differential Equations*. Cambridge University Press, 1996.

[Sch90] H.A. Schwarz. *Gesammelte Mathematische Abhandlungen*, volume 2. Springer, 1890. First published in Vierteljahrsschrift der Naturforschenden Gesellschaft Zürich, Vol. 15, 1870, pp. 272–286.

[SS98] P. Seshaiyer and M. Suri. Convergence results for non-conforming hp methods: The mortar finite element method. In J. Mandel, C. Farhat, and X. Cai, editors, *Domain Decomposition Methods 10, Boulder, August 1997*, pages 453–459. American Mathematical Society 218, 1998.

[Ste98] R. Stenberg. Mortaring by a method of J.A. Nitsche. In S. Idelsohn, E. Onate, and E. Dvorkin, editors, *Computational Mechanics: New Trends and Applications*, Barcelona, 1998. CIMNE.

[SZ90] L.R. Scott and S. Zhang. Finite element interpolation of nonsmooth functions satisfying boundary conditions. *Math. Comp.*, 54:483–493, 1990.

[Tos99] A. Toselli. *Domain Decomposition Methods for Vector Field Problems*. PhD thesis, Courant Institute of Mathematical Sciences, New York University, 1999.

[Tos00] A. Toselli. Overlapping Schwarz methods for Maxwell's equations in three dimensions. *Numer. Math.*, 86:733–752, 2000.

[TWW00] A. Toselli, O.B. Widlund, and B.I. Wohlmuth. Iterative substructuring method for Maxwell's equations in two dimensions. *Math. Comp.*, posted on March 1, 2000. PII: S 0025-5718(00)01244-8 (to appear in print).

[Ver84] R. Verfürth. A multilevel algorithm for mixed problems. *SIAM J. Numer. Anal.*, 21:264–271, 1984.

[Ver96] R. Verfürth. *A Review of A Posteriori Error Estimation and Adaptive Mesh-Refinement Techniques*. Wiley–Teubner, Chichester, 1996.

[Wid88] O.B. Widlund. Iterative substructuring methods: Algorithms and theory for elliptic problems in the plane. In R. Glowinski, G. Golub, G. Meurant, and J. Périaux, editors, *First International Symposium on Domain Decomposition Methods for Partial Differential Equations*, pages 113–128. SIAM, Philadelphia, 1988.

[Wid99] O.B. Widlund. Domain decomposition methods for elliptic partial differential equations. In H. Bulgak and C. Zenger, editors, *Error Control and*

Adaptivity in Scientific Computing, volume 536, pages 325–354. Kluwer Academic Publishers, 1999.

[WK99] B.I. Wohlmuth and R.H. Krause. Multigrid methods based on the unconstrained product space arising from mortar finite element discretizations. *Preprint A18-99, FU Berlin*, 1999. to appear in SIAM J. Numer. Anal.

[Woh95] B.I. Wohlmuth. *Adaptive Multilevel-Finite-Elemente Methoden zur Lösung elliptischer Randwertprobleme*. PhD thesis, TU München, 1995.

[Woh99a] B.I. Wohlmuth. Hierarchical a posteriori error estimators for mortar finite element methods with Lagrange multipliers. *SIAM J. Numer. Anal.*, 36:1636–1658, 1999.

[Woh99b] B.I. Wohlmuth. Mortar finite element methods for discontinuous coefficients. *Z. Angew. Math. Mech.*, 79 S I:151–154, 1999.

[Woh99c] B.I. Wohlmuth. A residual based estimator for mortar finite element discretizations. *Numer. Math.*, 84:143–171, 1999.

[Woh00a] B.I. Wohlmuth. A mortar finite element method using dual spaces for the Lagrange multiplier. *SIAM J. Numer. Anal.*, 38:989–1012, 2000.

[Woh00b] B.I. Wohlmuth. Multigrid methods for saddlepoint problems arising from mortar finite element discretizations. *ETNA*, 11:43–54, 2000.

[WTW00] B.I. Wohlmuth, A. Toselli, and O.B. Widlund. An iterative substructuring method for Raviart–Thomas vector fields in three dimensions. *SIAM J. Numer. Anal.*, 37:1657–1676, 2000.

[WW98] C. Wieners and B.I. Wohlmuth. The coupling of mixed and conforming finite element discretizations. In J. Mandel, C. Farhat, and X. Cai, editors, *Proceedings of the 10th International Conference on Domain Decomposition*, pages 546–553. AMS, Contemporary Mathematics series, 1998.

[WW99] C. Wieners and B.I. Wohlmuth. A general framework for multigrid methods for mortar finite elements. In W. Hackbusch and S. Sauter, editors, *Numerical Techniques for Composite Materials*, Notes on Numerical Fluid Mechanics. Vieweg, Braunschweig, Submitted to 15th GAMM–Seminar 1999. Preprint 415, Universität Augsburg, 1999.

[WY98] M.F. Wheeler and I. Yotov. Mortar mixed finite element approximations for elliptic and parabolic equation. In C.K. Chui et al., editor, *Approximation theory IX. Computational aspects. Proceedings of the 9th international conference*, volume 2, pages 377–392. Vanderbilt University Press, 1998.

[Xu92] J. Xu. Iterative methods by space decomposition and subspace correction. *SIAM Rev.*, 34:581–613, 1992.

[XZ98] J. Xu and J. Zou. Some nonoverlapping domain decomposition methods. *SIAM Rev.*, 40:857–914, 1998.

[Yot97] I. Yotov. A mixed finite element discretization on non-matching multiblock grids for a degenerate parabolic equation arising in porous media flow. *East-West J. Numer. Math.*, 5:211–230, 1997.

[Yse86] H. Yserentant. On the multi-level splitting of finite element spaces. *Numer. Math.*, 58:379–412, 1986.

[Yse93] H. Yserentant. Old and new proofs for multigrid methods. *Acta Numerica*, pages 285–326, 1993.

[Zul01] W. Zulehner. A class of smoothers for saddle point problems. *Computing*, to appear 2001. Institute of Analysis and Computational Mathematics, University of Linz, Austria, TR 546, 1998.

List of Figures

List of Tables

Notations

Basis functions:

ψ_i	basis function of $M_{h_m}(\gamma_m)$, $1 \leq i \leq N_m$
θ_i	basis function of $\widetilde{W}_{0;h_m}(\gamma_m)$, $1 \leq i \leq N_m$ and $W_{h_m}(\gamma_m)$, $1 \leq i \leq \nu_m$, $\nu_m > N_m$
$\tilde{\theta}_i$	basis function of $\widetilde{\widetilde{W}}_{h_m}(\gamma_m)$, $1 \leq i \leq N_m$
ϕ_i	standard nodal basis function

Domains and related notations:

D	Lipschitz domain
$|D|$	area of the domain D
Γ	interface or boundary
γ_m	interface between two subdomains
Ω	bounded polygonal domain
Ω_k	polygonal subdomain of Ω
$\Omega_{\bar{n}(m)}$	subdomain associated with the mortar side of the interface γ_m
$\Omega_{n(m)}$	subdomain associated with the non-mortar side of the interface
$\mathcal{S}$	union of all interfaces γ_m
$\mathcal{S}_m$	strip of width h_m on the mortar side of γ_m

Elasticity:

ϵ	infinitesimal strain tensor
$\mathbf{E}$	Hooke's tensor, (components E_{ijkl}, $1 \leq i, j, k, l \leq d$)
E, ν	Young's modulus, Poisson ratio
λ, μ	Lamé constants
σ	stress tensor, (components σ_{ij}, $1 \leq i, j \leq d$)
$\text{tr}(\cdot)$	trace of a tensor
$\mathbf{u}$, $\mathbf{f}$, $\mathbf{p}$	displacement field, volume force, surface traction

Finite element spaces:

B_{h_1}	space of cubic bubble functions on Ω_1
$B_h(\partial T)$	space of bubbles on ∂T
CR_{h_1}, CR_h	Crouzeix–Raviart finite element spaces on Ω_1, Ω
$M_{h_m}(\gamma_m)$	Lagrange multiplier space on the interface
M_h	Lagrange multiplier space
NC_{h_1}	enriched nonconforming finite element space on Ω_1
$NC_{g;h_1}$	subset of NC_{h_1} satisfying inhomogeneous boundary conditions
$\mathcal{ND}(\Omega; \mathcal{T})$	Nédélec finite elements in 3D associated with the triangulation $\mathcal{T}$
$\mathcal{ND}(T)$	local Nédélec finite elements in 3D

$Q(f)$	local space of bilinear functions on f
$RT_{h_1;n_1}$	Raviart–Thomas finite element space of order n_1 on Ω_1
$\mathcal{RT}(\Omega;\mathcal{T})$	Raviart–Thomas finite elements in 3D associated with $\mathcal{T}$
$\mathcal{RT}(T)$	local lowest order Raviart–Thomas finite elements in 3D
$S_h(\partial T)$	space of conforming bilinear finite elements on ∂T
V_h	constrained finite element space (Chapter 1)
V_h	finite element space on the fine triangulation (Chapter 2)
V_H	coarse finite element space associated with the macro-triangulation
V_l	finite element space on level l
$\widetilde{V}_l$	hierarchical basis finite element space on level l; $\widetilde{V}_l \subset V_l$
V_F	finite element space associated with two adjacent substructures
V_T	finite element space associated with one substructure
$W_{h_1;n_1}$	finite element space of piecewise polynomials of order n_1 on Ω_1
$W_{h_m}(\gamma_m)$	trace space of $X_{h_{n(m)};n_{n(m)}}$ on the interface γ_m
$W_{0;h_m}(\gamma_m)$	trace space with zero values on the boundary of γ_m
$\widetilde{W}_{h_m}(\gamma_m)$	subspace of $W_{h_m}(\gamma_m)$
$\widetilde{W}_{0;h_m}(\gamma_m)$	subspace of $W_{0;h_m}(\gamma_m)$
$X_{h_k;n_k}$	conforming P_{n_k}-finite elements on Ω_k
X_h	unconstrained product space

Finite element and weak solutions:

$\mathbf{j}$	flux of the weak solution
$\mathbf{j}_{h_1}$	discrete flux on Ω_1
λ	flux in normal direction
λ_h	discrete Lagrange multiplier
u	weak solution of the model problem
u_h	finite element solution

Hilbert spaces:

$C(D)$	continuous functions
$C_0(D)$	continuous functions with zero values on the boundary of D
$P_n(D)$	polynomials of degree $\leq n$
$H^s(D), L^2(D)$	standard Hilbert spaces
$H_0^1(D)$	subspace of $H^1(D)$ with zero trace on the boundary of D
$H_{00}^{\frac{1}{2}}(\gamma_m)$	interpolation space between $L^2(\gamma_m)$ and $H_0^1(\gamma_m)$
$\left(H_{00}^{\frac{1}{2}}(\gamma_m)\right)'$	dual space of $H_{00}^{\frac{1}{2}}(\gamma_m)$
$H(\mathrm{div};D)$	vector valued Hilbert space; $\mathrm{div}\,\mathbf{q} \in L^2(D)$
$H_0(\mathrm{div};D)$	subspace of $H(\mathrm{div};D)$; $\mathbf{q}\cdot\mathbf{n} = 0$ on ∂D
$H(\mathbf{curl};D)$	vector valued Hilbert space; $\mathbf{curl}\,\mathbf{q} \in (L^2(D))^3$
Y	subspace of $\prod_{k=1}^K H^1(\Omega_k)$; $\int_{\gamma_m}[v]\,d\sigma = 0$, $1 \leq m \leq M$

Miscellaneous:

δ_{ij}	Kronecker symbol
η_T	local contribution of an error estimator
$\hat{\eta}_T$	local contribution of a simplified error estimator
K	number of subdomains
κ	condition number
λ	eigenvalue in Subsect. 2.3.3
M	number of interfaces

N_m	dimension of the Lagrange multiplier space $M_{h_m}(\gamma_m)$
ν_m	dimension of the trace space $W_{h_m}(\gamma_m)$
ρ	spectral radius
σ	spectrum of an operator
χ	performance
ζ	effectivity index

Norms and semi norms:

$\|\cdot\|$	Euclidean vector norm
$\|\cdot\|_0$	L^2-norm on Ω
$\|\cdot\|_1$	broken H^1-norm on Ω
$\|\cdot\|_1$	broken H^1-semi norm on Ω
$\|\!\|\cdot\|\!\|$	energy norm
$\|\cdot\|_{s;D}$	H^s-semi norm on D
$\|\cdot\|_{s;D}$	H^s-norm on D (scaled or unscaled)
$\|\cdot\|_{s;\partial D}$	H^s-semi norm on ∂D
$\|\cdot\|_{s;\partial D}$	H^s-norm on ∂D (scaled or unscaled)
$\|\cdot\|_{H^{\frac{1}{2}}(\gamma_m)}$	$H^{1/2}$-norm on γ_m
$\|\cdot\|_{H^{\frac{1}{2}}_{00}(\gamma_m)}$	$H^{1/2}_{00}$-norm on γ_m
$\|\cdot\|_{(H^{\frac{1}{2}}_{00}(\gamma_m))'}$	$H^{1/2}_{00}$-dual norm on γ_m
$\|\cdot\|_{h^s_m;\gamma_m}$	mesh dependent L^2-norm on γ_m (local factor h_e^{-s})
$\|\cdot\|_{h^s_m;S}$	mesh dependent L^2-norm on S (local factor h_e^{-s})
$\|\cdot\|_{M_m}$	Lagrange multiplier norm on γ_m
$\|\cdot\|_{M'_m}$	dual norm of $\|\cdot\|_{M_m}$
$\|\cdot\|_M$	Lagrange multiplier norm on S
$\|\cdot\|_{M'}$	dual norm of $\|\cdot\|_M$

Operators:

I_k	Lagrange interpolation operator on Ω_k
Π_m	mortar projection
P_m	projection onto $\widetilde{W}_{h_m}(\gamma_m)$
Q_m	dual projection onto $M_{h_m}(\gamma_m)$
ρ_H, ρ_l	interpolation operator for Raviart–Thomas vector fields

Triangulations and related notations:

h_{mor}	meshsize on the mortar side
h_{non}	meshsize on the non-mortar side
h_e	length of the edge e (2D), diameter of the element e (3D)
H/h	ratio between substructure diameter and element diameter
$\mathcal{E}_h, \mathcal{E}_H$	sets of edges
E, e	edges of the triangulation
$\mathcal{F}_h, \mathcal{F}_H$	sets of faces
F, f	faces of the triangulation (in 3D)
$S_{m;h_m}$	triangulation on the interface γ_m (non-mortar side)
$S_{m;l}$	family of triangulations on the interface γ_m
$\mathcal{T}_{k;h_k}$	triangulation on Ω_k with meshsize h_k
$\mathcal{T}_h, \mathcal{T}_H$	global triangulations
$\mathcal{T}_l$	family of shape regular triangulations
T, t	elements of the triangulation

Index

Editorial Policy

§1. Volumes in the following four categories will be published in LNCSE:

i)   Research monographs
ii)   Lecture and seminar notes
iii) Conference proceedings
iv)   Textbooks

Those considering a book which might be suitable for the series are strongly advised to contact the publisher or the series editors at an early stage.

§2. Categories i) and ii). These categories will be emphasized by Lecture Notes in Computational Science and Engineering. **Submissions by interdisciplinary teams of authors are encouraged.** The goal is to report new developments – quickly, informally, and in a way that will make them accessible to non-specialists. In the evaluation of submissions timeliness of the work is an important criterion. Texts should be well-rounded, well-written and reasonably self-contained. In most cases the work will contain results of others as well as those of the author(s). In each case the author(s) should provide sufficient motivation, examples, and applications. In this respect, Ph.D. theses will usually be deemed unsuitable for the Lecture Notes series. Proposals for volumes in these categories should be submitted either to one of the series editors or to Springer-Verlag, Heidelberg, and will be refereed. A provisional judgment on the acceptability of a project can be based on partial information about the work: a detailed outline describing the contents of each chapter, the estimated length, a bibliography, and one or two sample chapters – or a first draft. A final decision whether to accept will rest on an evaluation of the completed work which should include

– at least 100 pages of text;
– a table of contents;
– an informative introduction perhaps with some historical remarks which should be accessible to readers unfamiliar with the topic treated;
– a subject index.

§3. Category iii). Conference proceedings will be considered for publication provided that they are both of exceptional interest and devoted to a single topic. One (or more) expert participants will act as the scientific editor(s) of the volume. They select the papers which are suitable for inclusion and have them individually refereed as for a journal. Papers not closely related to the central topic are to be excluded. Organizers should contact Lecture Notes in Computational Science and Engineering at the planning stage.

In exceptional cases some other multi-author-volumes may be considered in this category.

§4. Category iv) Textbooks on topics in the field of computational science and engineering will be considered. They should be written for courses in CSE education. Both graduate and undergraduate level are appropriate. Multidisciplinary topics are especially welcome.

§5. Format. Only works in English are considered. They should be submitted in camera-ready form according to Springer-Verlag's specifications. Electronic material can be included if appropriate. Please contact the publisher. Technical instructions and/or TeX macros are available via http://www.springer.de/author/tex/help-tex.html; the name of the macro package is "LNCSE – LaTeX2e class for Lecture Notes in Computational Science and Engineering". The macros can also be sent on request.

General Remarks

Lecture Notes are printed by photo-offset from the master-copy delivered in camera-ready form by the authors. For this purpose Springer-Verlag provides technical instructions for the preparation of manuscripts. See also *Editorial Policy*.

Careful preparation of manuscripts will help keep production time short and ensure a satisfactory appearance of the finished book. The actual production of a Lecture Notes volume normally takes approximately 12 weeks.

The following terms and conditions hold:

Categories i), ii), and iii):
Authors receive 50 free copies of their book. No royalty is paid. Commitment to publish is made by letter of intent rather than by signing a formal contract. Springer-Verlag secures the copyright for each volume.

For conference proceedings, editors receive a total of 50 free copies of their volume for distribution to the contributing authors.

Category iv):
Regarding free copies and royalties, the standard terms for Springer mathematics monographs and textbooks hold. Please write to Peters@springer.de for details. The standard contracts are used for publishing agreements.

All categories:
Authors are entitled to purchase further copies of their book and other Springer mathematics books for their personal use, at a discount of 33,3 % directly from Springer-Verlag.

Addresses:

Professor M. Griebel
Institut für Angewandte Mathematik
der Universität Bonn
Wegelerstr. 6
D-53115 Bonn, Germany
e-mail: griebel@iam.uni-bonn.de

Professor D. E. Keyes
Computer Science Department
Old Dominion University
Norfolk, VA 23529–0162, USA
e-mail: keyes@cs.odu.edu

Professor R. M. Nieminen
Laboratory of Physics
Helsinki University of Technology
02150 Espoo, Finland
e-mail: rni@fyslab.hut.fi

Professor D. Roose
Department of Computer Science
Katholieke Universiteit Leuven
Celestijnenlaan 200A
3001 Leuven-Heverlee, Belgium
e-mail: dirk.roose@cs.kuleuven.ac.be

Professor T. Schlick
Department of Chemistry and
Courant Institute of Mathematical
Sciences
New York University
and Howard Hughes Medical Institute
251 Mercer Street, Rm 509
New York, NY 10012-1548, USA
e-mail: schlick@nyu.edu

Springer-Verlag, Mathematics Editorial
Tiergartenstrasse 17
D-69121 Heidelberg, Germany
Tel.: *49 (6221) 487-185
e-mail: peters@springer.de
http://www.springer.de/math/
peters.html

Lecture Notes in Computational Science and Engineering

For further information on these books please have a look at our mathematics catalogue at the following URL: http://www.springer.de/math/index.html